2nd International Conference on Design and Manufacturing Engineering (ICDME 2017)

IOP Conference Series: Materials Science and Engineering
Volume 239

Guangzhou, China
1 – 3 August 2017

ISBN: 978-1-5108-4998-3
ISSN: 1757-8981

Printed from e-media with permission by:

Curran Associates, Inc.
57 Morehouse Lane
Red Hook, NY 12571

CURRAN ASSOCIATES INC.
proceedings
.com

Some format issues inherent in the e-media version may also appear in this print version.

Printed by Curran Associates, Inc. (2017)

For permission requests, please contact the Institute of Physics
at the address below.

Institute of Physics
Dirac House, Temple Back
Bristol BS1 6BE UK

Phone: 44 1 17 929 7481
Fax: 44 1 17 920 0979

techtracking@iop.org

Additional copies of this publication are available from:

Curran Associates, Inc.
57 Morehouse Lane
Red Hook, NY 12571 USA
Phone: 845-758-0400
Fax: 845-758-2633
Email: curran@proceedings.com
Web: www.proceedings.com

Table of contents

Volume 239

2nd International Conference on Design and Manufacturing Engineering (ICDME2017)

1–3 August 2017, Guangzhou, China

Accepted papers received: 7 September 2017
Published online: 16 October 2017

Fluid mechanics analysis and Application

Material performance analysis and test

IOP Conf. Series: Materials Science and Engineering **239** (2017) 011001 doi:10.1088/1757-899X/239/1/011001

PREFACE

The 2017 International Conference on Design and Manufacturing Engineering (ICDME 2017) was successfully held at Guangdong University of Technology, Guangzhou, a famous megacity for remarkable history and fine cuisine, China from August 1-3, 2017. Guangzhou also has the most dynamic and pioneering economy with fast growing manufacturing sector spearheaded by automotive and shipbuilding industries, among many others, demanding the latest knowledge and technology for creative products and efficient manufacturing process in meeting challenges of future manufacturing technology. The ICDME 2017 reflected the technology needs of Guangzhou and beyond with its global participation and diverse but focused topics concerning design and manufacturing engineering presented in the truly international conference. This proceeding contains the reviewed papers presented at the ICDME 2017 and covered most hot and important issues related to the design and manufacturing engineering of many industries with a focus on the automotive technology.

The ICDME 2017 conference program consisted of presentations in forms of keynote, oral, and poster from researchers, engineers, and graduate students working in fields of materials science, mechanical engineering, measurement technology, materials processing, and design methodology to report novel methods, findings, and designs. Although in its second since the inception, the ICDME has positioned itself as a platform for technical exchanges, academic promotion, and collaboration enabling for participants with diverse interests, background, objectives, and expertise. With the fast globalization of technology and manufacturing, technical issues must be discussed and disseminated in a global forum participated by researchers and engineers with international experiences and activities. This goal of the ICDEM is well achieved by the international participants from countries with extensive technology development such as Egypt, Malaysia, Turkey, in addition to more active countries like Germany, China, Canada, India, and Australia.

As its tradition, in this year's ICDME, there were keynote talks presented by internationally renowned scholars from top institutions with broad topics on design and manufacturing. Then there were sessions focused on materials, signal analysis, vehicular engineering, machine structures, mechatronics, and robotics with oral presentations and posters. Papers in this volume are selections based on reviews from the conference technical committee and they reflected the high quality and broad technical interests of the conference.

We want to express our gratitude to all members of conference committees and reviewers who spent their valuable time for the successful conference with carefully organized programs, well-planned sessions, high quality conference papers, and pleasant social hours and events. We would also like to thank all authors who have contributed to this conference and committee members, reviewers, speakers, session chairs, sponsors, and support staff for the great success of ICDME 2017.

Professor Ji Wang

Chair, Scientific and Publication Committee

August 11, 2017

IOP Conf. Series: Materials Science and Engineering **239** (2017) 011002 doi:10.1088/1757-899X/239/1/011002

CONFERENCE COMMITTEE CHAIRS

Prof. Dr. Ridha Ben Mrad, University of Toronto, Canada
Prof. Dennis Y.C. Leung, The University of Hong Kong, Hong Kong
Prof. John Mo, Royal Melbourne Institute of Technology, Australia
Prof. Ji Wang, Ningbo University, China

CONFERENCE CONFERENCE CO-CHAIR

Prof. Feng Wang, Guangdong University of Technology, China

PROGRAM COMMITTEE CHAIRS

Assoc. Prof. JING Xingjian, The Hong Kong Polytechnic University, Hong Kong
Prof. M. Chandrasekaran, Vels University, India
Prof.Yu-Liang Chen, Chung Cheng Institute of Technology, National Defense University, Taiwan

PUBLICITY CHAIR

Prof. M. Chandrasekaran, Vels University, India

INTERNATIONAL TECHNICAL COMMITTEES

Prof. Usman Sudjadi, National Nuclear Energy Agency, Indonesia
Prof. Ming-Liang Liao, Air Force Institute of Technology, Taiwan
Prof. Long-Chang Hsieh, National Formosa University, Taiwan
Prof. Dr. Cho-Pei Jiang, National Formosa University, Taiwan
Prof. Shubhashis Sanyal, NIT Raipur, India
Prof. T.R.Vijayaram, School of Mechanical and Building Sciences, VIT University, Chennai, India
Assoc. Prof. Zhen Liu, South China University of Technology (SCUT), China
Assoc. Prof. Dai-Jia Juan, National Formosa University, Taiwan
Assoc. Prof. Syh-Shiuh Yeh, Dept. Mechanical Engineering, National Taipei University of Technology, Taiwan
Assoc. Prof. Zhang Shi-jun, Shandong Transport Vocational College, China
Assoc. Prof. Dr.M. K. Bhatt, S. V. National Institute of Technology, India
Assoc. Prof. Dr.Yeong-Sant Kuo, ChienKuo Technology University, Taiwan
Assoc. Prof. Chel-Jong Choi, Chonbuk National University, Republic of Korea
Assoc. Prof. Yao-Wen Tsai, Da-Yeh University, Taiwan R.O.C
Assoc. Prof. En-Chih Chang, I-Shou University, Taiwan, R.O.C.
Assoc. Prof. Yu-Ren Wu, National Central University, Taiwan
Assoc. Prof. Yung-Chang Cheng, National Kaohsiung First University of Science and Technology, Taiwan

IOP Conf. Series: Materials Science and Engineering **239** (2017) 011002 doi:10.1088/1757-899X/239/1/011002

IOP Conf. Series: Materials Science and Engineering **239** (2017) 011003 doi:10.1088/1757-899X/239/1/011003

Peer review statement

All papers published in this volume of *IOP Conference Series: Materials Science and Engineering* have been peer reviewed through processes administered by the proceedings Editors. Reviews were conducted by expert referees to the professional and scientific standards expected of a proceedings journal published by IOP Publishing.

IOP Conf. Series: Materials Science and Engineering **239** (2017) 012001 doi:10.1088/1757-899X/239/1/012001

EHPS Handling Stability Analysis of Electric Bus Based on System Identification Method

Ni Zhang, Liu Hai-mei , Bei Shao-yi, Li Bo, Zhao Jing-bo

School of Automobile and Traffic Engineering, Jiangsu University of Technology, Changzhou 213001, China
119443085@qq.com

Abstract. Electric hydraulic assist force steering system (EHPS system) is the steering system of electric bus, this paper presents a method of EHPS handling stability analysis based on system identification method according to the handling stability of EHPS system for electric bus.The simulation model of electro-hydraulic assist force steering system EHPS is established by using the software AMESim, and making a quantitative analysis on the characteristics of the electric assist force assisted steering system, the assist force response and stability. At the same time, we study the stability of vehicle, including hunting, transient response, return experiment, the results show that the HPS and EHPS by comparing the simulation: It improves the portability, road sense, transient response and return performance after loading the system, which verifiy the effectiveness of the control strategy that improves vehicle steering performance, and it provides the basis for the optimization of control methods in the future.
Keywords: EHPS; handling stability analysis; AMESim; control strategy; electric bus

1. Introduction

With the development of system identification methods, the method of identification modeling has been widely used in the field of vehicle engineering. It proved to be effective that the off-line identification method based on neural network and the online identification method based on Delft tire model in the identification of road friction coefficient. It is necessary to use the software AMESim to establish the simulation model of electro-hydraulic assist force steering system EHPS, which makes a quantitative analysis of the characteristics of assist force assist, assist force response and stability of the electric hydraulic assist force steering system, The vehicle handling stability performance is studied to verify the effectiveness of the control strategy for improving vehicle steering performance, and which provide the basis for the optimization of control methods in the future.

2. EHPS system model based on AMESim

The input signal simulation model is shown in Figure 1.

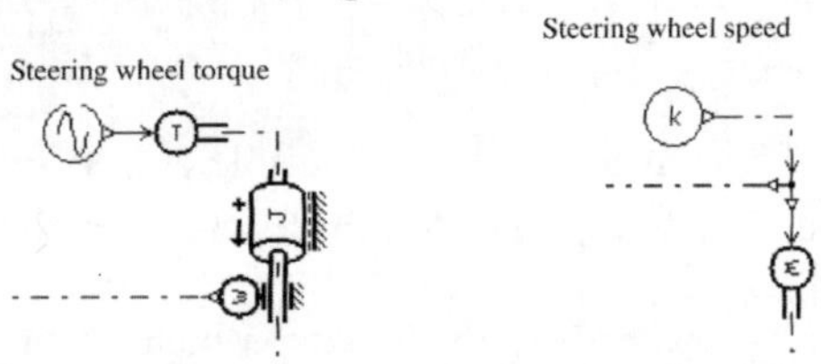

Figure 1. Simulation Module of Input Signal

IOP Conf. Series: Materials Science and Engineering **239** (2017) 012001 doi:10.1088/1757-899X/239/1/012001

When the driver steers the steering wheel, the steering wheel can be viewed in two ways, angle input and force input. When the vehicle is at low speed, the input port is used, and the force input port is used at high speed.

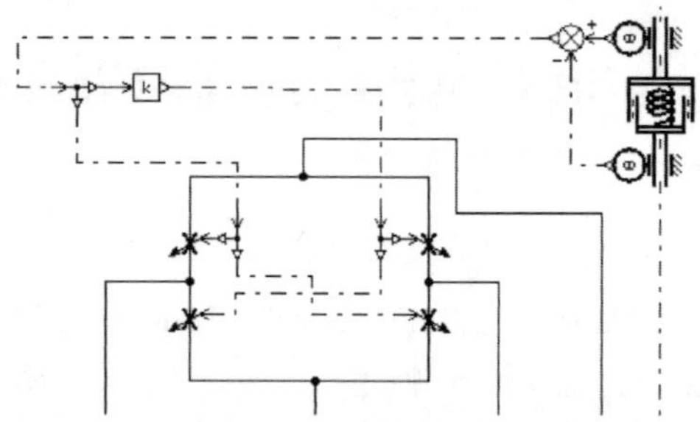

Figure 2. Simulation Module of Rotation Valve

The model of Rotation Valve is shown in Figure 2. Torsional spring simulation module with proper stiffness is used to simulate the torsion bar. The upper and lower ends of the torsion bar are respectively provided with an angular displacement signal output module, the difference between the output signals of the two modules is used as the deformation signal of the torsion bar. The signal is then transferred to the model of rotation valve.

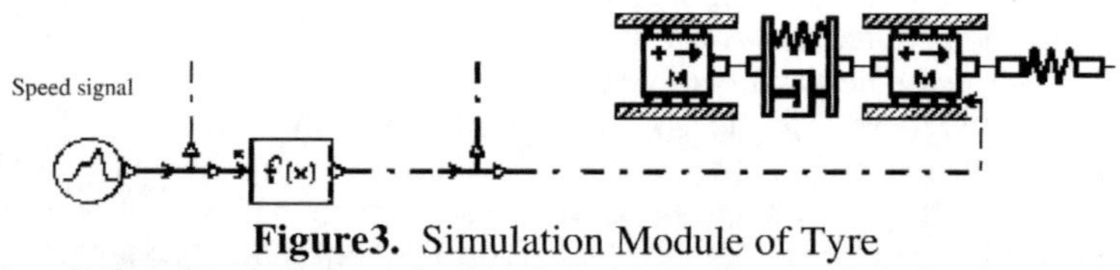

Figure3. Simulation Module of Tyre

In the simulation, the sliding friction force in the tire simulation module is changed according to the formula, So we can simulate the change of the driving resistance during the driving process, as shown in Figure 3.

The control model of the motor is established according to the dynamic mathematical model of the motor, and the requirement of motor control is accurate and quick response.

The steering resistance is different when the vehicle is traveling at different speeds, the current is different because the motor current is linear with the load, the voltage drop of the armature resistance varies with the current, so that the back EMF is different at different speeds. That is to say, the relationship between the motor voltage and the motor speed is not constant but with the speed change, so in order to meet the requirements of motor speed control, the use of the speed loop closed loop control method.

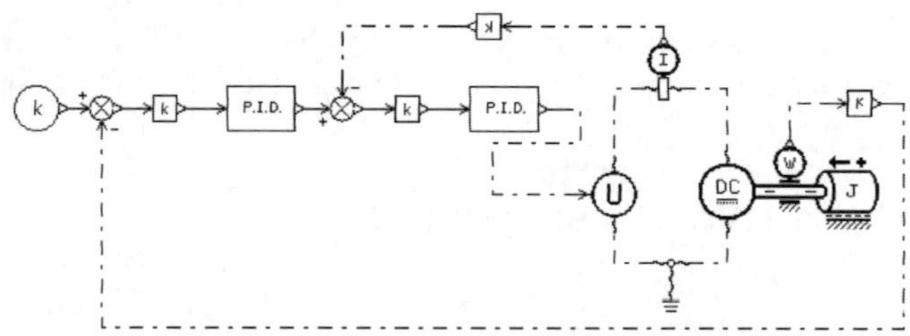

Figure 4. Simulation Module of double closed-loop motor control

Control of the motor using the speed loop and current loop that called double closed loop PID speed control system, as shown in Figure 4.

The motor speed is controlled by controlling the motor voltage through using the DC motor to drive the hydraulic pump, it can limit the maximum current to prevent the gear pump gear being broken and it also avoid the motor being damaged when the current is surged. At the same time, the motor speed feedback signal and the current feedback signal are analyzed to determine the control voltage.

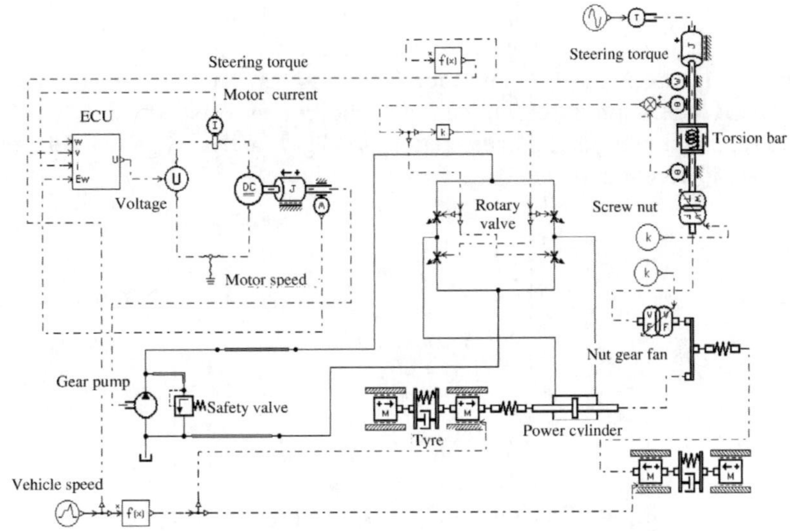

Figure 5. Simulation Module of EHPS of Bus

The simulation model of the whole EHPS system is shown in Figure 5, and the simulation model of the whole bus EHPS system is set up through reasonably make each module connected .

3. EHPS system handling stability analysis

In order to improve the steering performance of the EHPS system compared with the HPS system, the two steering systems are simulated and the simulation results are compared.

3.1 Simulation and analysis of assist force response characteristics

Assist force response characteristics include two aspects: assist force response speed and stability.

Due to the existence of the valve structure, the bus EHPS and HPS system can be based on the mechanical structure to enter the booster cylinder hydraulic oil flow changes according to the change of the steering wheel speed. Therefore, we only need to verify the control strategy of the bus EHPS system, in which the control adjustment that the relationship between assist force size and steering wheel speed is appropriate or not.

We set the speed of 50km/h in the simulation, respectively, with 120deg/s, 360 deg/s, 500 deg/s, 720 deg/s, 800 deg/s rotating speed uniform steering wheel. And we get the curve of simulation of hydraulic assist force in various steering wheel speed change , that is, the assist force response curve, as shown in Figure 6.

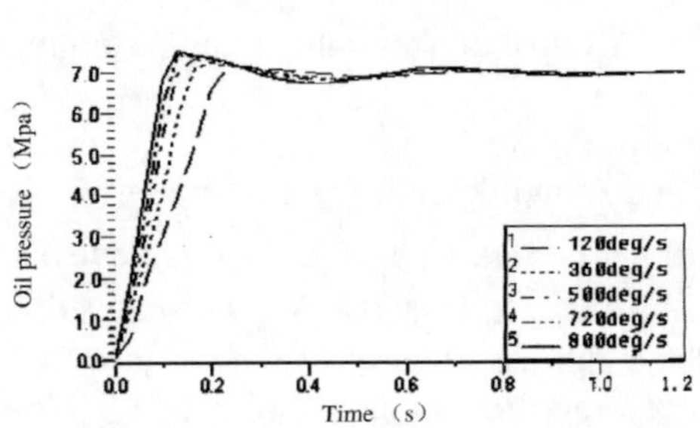

Figure 6. Assist Force on Rack with Various Steering Speed

In response time (the first time of reach steady state) and overshoot (maximum assist force / steady state) indicate the assist force response and stability. As shown in Figure 6, hit the steering wheel, the response time is less than 0.3s, and the maximum overshoot does not exceed 10% of the steady-state value. It shows that the assist force response and stability are good.

3.2 Simulation and analysis of the middle position handling stability

Analyzing the high speed steering and steering wheel force through the simulation of snake working condition. The maximum speed of 70%, and rounding an integer multiple of 10, in the simulation, the model keeps the speed of 90km/h in serpentine road condition as a sinusoidal curve. The period of sine motion is 5S, and the maximum lateral acceleration is 0.2g. Finally, we get simulation of the steering wheel torque and lateral acceleration curve, as shown in Figure 7.

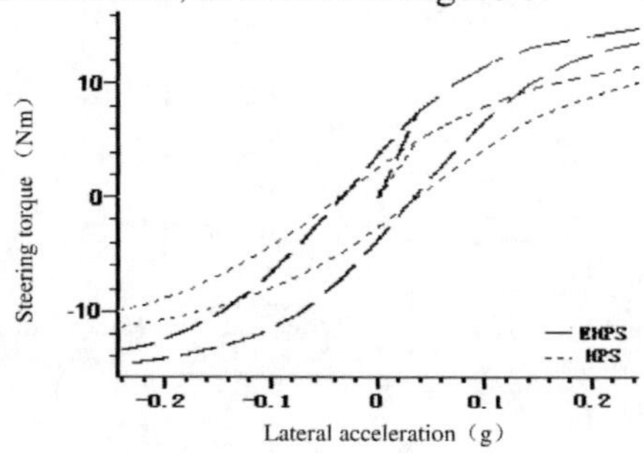

Figure 7. Handwheel Torque vs Lateral Acceleration

3.3 Simulation analysis of transient response

Simulation and analysis of vehicle transient response performance is based on steering wheel angle step input. The maximum speed of 70%, and rounding an integer multiple of 10, in the simulation, the model is driven by 90km/h speed. The steering wheel is applied to a ramp angle input of the 20 rad/s to produce a 30 degree angle, and the vehicle yaw rate variation curve is shown in figure 8.

The simulation results are shown in table 1, the reaction time is called as the time when the yaw rate of the vehicle reaches the steady yaw rate for the first time. We can see the result through comparing simulation of EHPS and HPS. The vehicle loaded with EHPS has a shorter reaction time, and the driver feels more responsive; The steady yaw rate is small, which increases the vehicle understeer.

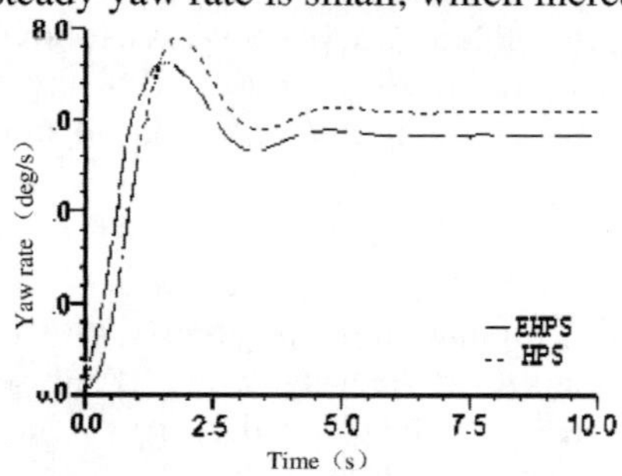

Figure 8. Yaw-rate of angle step input

3.4 Simulation and analysis of return performance

The maximum speed of 70%, and rounding an integer multiple of 10, in the simulation, the model is driven by 90km/h speed. Turning the steering wheel to achieve lateral acceleration of 20 m/s^2, loosen the steering wheel suddenly after it is stable. We set the steering wheel torque to zero through in the simulation. Recording 4 seconds after the release of the movement process. Then we analyse the relationship between yaw velocity and time, as shown in figure 9. Compared with the EHPS and HPS residual yaw rate can be seen, after loading EHPS, the performance of high speed recovery is significantly improved, and which is near completely return-to -center.

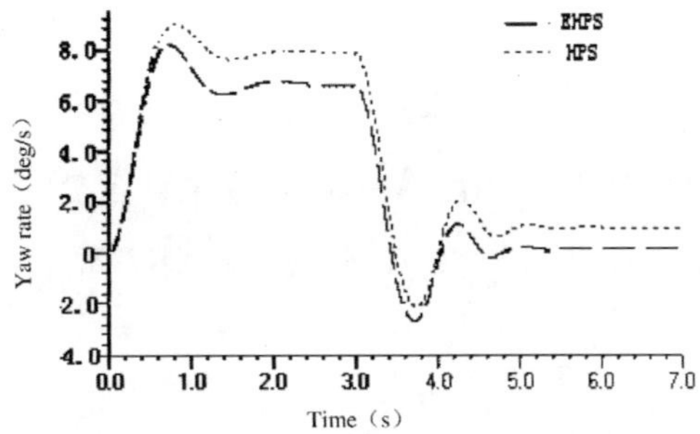

Figure 9. Yaw-rate of return-to-center

4. Conclusion

The simulation model of electro-hydraulic power steering system EHPS is established by using the software AMESim, and the paper makes a quantitative analysis of the assist force characteristics, assist force response and stability of the electric hydraulic power steering system, we also research the stability of vehicle, including hunting, transient response, return-to-center experiment, the results show that the system can improve the portability, road sense, transient response and return-to-return performance by comparing the simulation results of HPS and EHPS, which verify the effectiveness of the control strategy in improving the steering performance of the vehicle and provides the basis for the optimization of the control method in the future.

Acknowledgment

Funding from the National Natural Science Foundation of China (Grant No: 61503163), the "333 project" of Jiangsu Province (Grant No: BRA2016440) and the six talent peaks project in Jiangsu Province (Grant No: ZBZZ-024) are gratefully acknowledged, Changzhou Science and Technology Support Plan Project(CE20150084), Ministry of Housing and Urban-Rural Science and Technology Program(KYZB16002).

References

[1] Keld Laursen, Toke Reichstein and Ammon Salter 2011 *J. Regional Studies* **45**(4) 507-23
[2] Qi X, Wu G and Boriboonsomsin K 2016 *J. Transportation Research Record Journal of the* Transportation *Research Board* (2572) 1-8
[3] Kim N, Cha S and Peng H 2011 *J. IEEE Transactions on Control Systems Technology* **19**(5) 1279-87
[4] Zhu X, Zhang H and Cao D 2015 *J. Mechanical Systems & Signal Processing* (s58–59) 15-28
[5] Rahman K M, Jurkovic S and Stancu C 2012 *J. IEEE Transactions on Industry Applications* **51**(3) 2479-88
[6] Qiu X, Huang W and Bu F 2013 *J. Journal of Power Electronics* **13**(6) 964-74
[7] Zhu X, Zhang H and Fang Z 2015 *J. Mechanical Systems & Signal Processing* (s64–65) 46-57
[8] Zhu D, Zhou D and Zhou J 2012 *J. Journal of Dynamic Systems Measurement & Control* **134**(4) 041007
[9] Budde-Meiwes H, Drillkens J and Lunz B 2013 *J. Proceedings of the Institution of Mechanical Engineers Part D Journal of Automobile Engineering* **227**(5) 61-776
[10] Anwar M, Hayes M and Tata A 2015 *J. Lecture Notes in Computer Science* **4**(1) 94-103
[11] Sciarretta A, Serrao L and Dewangan P C 2014 *J. Control Engineering Practice* **29**(6) 287-98
[12] Qian L, Qiu L and Xin F 2014 *J. Transactions of the Chinese Society of Agricultural Engineering* **30**(19) 55-64

IOP Conf. Series: Materials Science and Engineering **239** (2017) 012002 doi:10.1088/1757-899X/239/1/012002

Automatic Tool Selection in V-bending Processes by Using an Intelligent Collision Detection Algorithm

A A Salem

Department of Mechanical Design and Production, Faculty of Engineering, Cairo University, Giza, Egypt.

amrabdel-alim@edu.cu.eg

Abstract. V-bending is widely used to produce the sheet metal components. There are global Changes in the shape of the sheet metal component during progressive bending processes. Accordingly, collisions may be occurred between part and tool during bending. Collision-free is considered one of the feasibility conditions of V-bending process planning which the tool selection is verified by the absence of the collisions. This paper proposes an intelligent collision detection algorithm which has the ability to distinguish between 2D bent parts and the other bent parts. Due to this ability, 2D and 3D collision detection subroutines have been developed in the proposed algorithm. This division of algorithm's subroutines could reduce the computational operations during collisions detecting.

1. Introduction

Industrial companies require quick products with high quality levels and low costs to meet the high levels of competition in today's market demand. This need demands efficient product's process plans which reduce the production time by minimizing the non-added-value activities. Correspondingly, the subtle bending tool selection is one of the goals of the efficient V-bending process planning. Collisions may be occurred during the bending process due to the global changes of the shape of the sheet metal and these collisions are considered as a one of the infeasibility conditions of the V-bending process planning. Consequently, to avoid these collisions, the collision detection algorithm must be followed.

Much of the literature on the V-bending process planning have detected the collisions by using external collision detection library or by developing a collision detection algorithm. The collision detection algorithms are divided into three generic categories [1]. In the first one, collisions are detected in situations between numbers of convex polyhedral. Meanwhile, space occupancy method has been used in the second category. Besides, the sweep volume calculation method is used between the moving objects in last category. In [2] and [3], collisions are detected in certain situations (before and after the bending process) of the bent parts which have only parallel bend lines. Consequently, collisions-free bending process verifies the selection of the bending tool in a certain process plan. The Space occupancy method is used to detect the collisions in [4] by subtracting the volume of the bent part of the volume of the tools, and then, if the volume of tool is divided into two solid volumes, the bent part could collide the tool. Geometrical constraints have developed in [5] to avoid collisions between the part and tools and these constraints are considered a pre-selection stage of the tool selection algorithm to minimize the change of tools during bending processes of one part. On contrary to [5], geometrical constraints have used in [6] to design the collision-free bending tool for a group of bent parts. However, much collision detection algorithms have been developed, they don't pay much

attention to the relations between bend lines and the benefits of these relations on controlling the computational loads of the detecting process.
The proposed intelligent collision detection algorithm doesn't detect only the collisions during bending the bent parts which have only parallel bend lines, but also detect the collisions of the bent parts which have bend lines in different directions. Moreover, it has ability to recognize the relations between the bend lines in sheet metal part. Accordingly, there are two subroutines in this algorithm which are 2D collision detection subroutine for the bent parts which have only parallel bend lines (2D bent parts) and 3D collision detection subroutine for the other bent parts (3D bent parts). This division could reduce the computational loads by detecting collisions of 2D bent parts without need of performing much computational operations as in 3D collision detection subroutine. Besides, the proposed algorithm uses simplified intersection checks which depend on the parametric equations.

2. Tool selection in V-bending

The V-bending process is carried out on the press brake by using its movable and static parts where the punch press down the sheet metal to the die with enough tonnage. There are different types of V-bending tools such as sash, gooseneck, and narrow relief punches and different V-dies. Accordingly, to select a punch and die to perform V-bending of certain bend line, the tool must satisfy the technological constraints as described in [5]. After satisfying the technological constraints, these tools must lead to collision-free bending process. Due to the interference between the bending sequence and the tool selection of V-bending process, a collision could be occurred by using a specific bending tool set and could not be occurred by using another tool. For instance, the collision detection process is performed in V-bending processes on bent part which shown in figure 1(a) by using two different bending punches which shown in figure 1(b). Even though there is a collision during bending b2 by using first punch according to the bending sequence {b1, b3, b2} as shown in figure 1(c), there aren't collisions by using the second punch according to the same sequence as shown in figure 1(d).

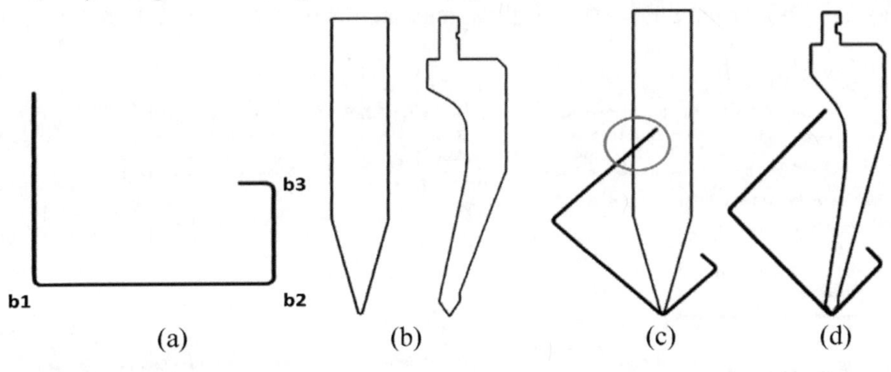

Figure 1. (a) sheet metal part, (b) two different bending punches, (c) collision in bending b2, (d) collision-free in bending b2.

3. The proposed collision detection algorithm

The proposed collision detection algorithm is developed using the C++ programming language under Windows operating system. This algorithm depends on detecting the collisions during static situations of the bending processes. The following flowchart which shown in figure 2 presents the operations of the proposed algorithm. As presented the first operation is the data extraction of the bent part from STEP-AP203 format file, and then the extracted geometrical data of the bent part are the input data to the feature recognition and reasoning system as described in [7]. This feature recognition system produces the central surface of the flat pattern of the part which is used in collision detection algorithm. The algorithm proposes bending sequence and tool list to perform bending of the part where the feasibility of proposed sequence and tool list is tested by the proposed collision detection algorithm. Bend lines are bent one by one due to the bending sequence and the algorithm detects the collisions of each bending process.

The bending of the bend line includes two operations, which are the transformation of bend line from the planar surface (flat pattern of the bend line) to the cylindrical surface and the rotation of the two portions of the bent part which adjacent to this bend line by a half of the bending angle. After bending operation is finished, the setup operation is started. The setup operation is the operation of the positioning of the bent part and tool set as in real bending. Then, the collisions are detected by 2D or 3D collision detection subroutine. As soon as the bent part is classified as 2D bent part, collisions are detected by 2D subroutine. Otherwise, collisions are detected by 3D subroutine.

3.1. 2D collision detection subroutine

As the 2D bent part could be represented by its projection in 2D plane, the collisions between 2D bent parts and bending tool could be presented in 2D projection of this bending process. On the contrary to 2D bent parts, the bending of 3D bent parts couldn't represented by the projection in 2D plane, because collision between faces which don't align to bending tool axis and bending tool couldn't be represented in 2D plane projection. Due to the projections of part and bending tools consist of line and arc edges, the collisions between tools and bent part are the intersections between any line of bent part projection and any line or arc of bending tool projection as shown in figure 1(c). Firstly, this subroutine extracts the projections of bending tools and part from the geometrical data. These projections consist of any edge of tool and part which doesn't align to the tool axis. After the projections extracting, the detection process for line-line intersection or line-circle (arc) intersection will start.

In line-line intersection, there are only two relations between lines in 2D plane, the first one is parallelism relation and the second one is intersection relation. If the slopes of two lines are equal to each other, these lines are parallel lines and couldn't intersect with each other. Unless the slopes of lines are equal, these lines are intersecting lines. The intersection point is the point which could satisfy the equation line of two intersecting lines. In line-circle (arc) intersection, the 2D collision detection subroutine uses the perpendicular distance between the center of the circle (the circle of this arc) and the line. If this distance is shorter than the radius of the circle, there would be intersection between the line and the circle.

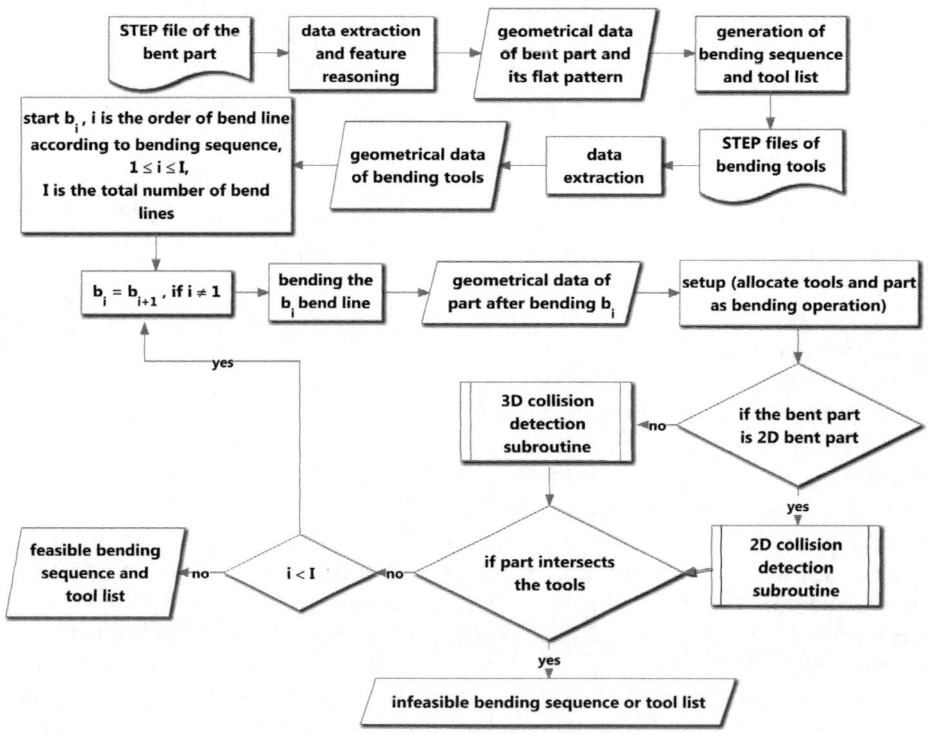

Figure 2. flowchart of the intelligent collision detection algorithm

3.2. 3D collision detection subroutine

If the bent part collides the tool set on the press brake, at least one face (planar polygon) of this part intersects with one polygon (planar surface or cylindrical surface) of the tool set. Consequently, to detect the collisions, intersection between two polygons in 3D must be detected. As soon as there is a collision between two polygons in 3D, at least one edge of one polygon must intersect the other polygon and versa vice as shown in figure 3. Parametric equation of line edge, planar surface, and cylindrical surface, which are presented in equations (1), (2), (3) respectively, are used to determine the intersection point. Where (x_s, y_s, z_s) and (x_e, y_e, z_e) in equation (1) are the coordinates of the end points of line edge, and the parameter t represents the position of any point lies between the end points, and (a, b, c, d) in equation (2) are determined from normal vector to the planar surface and one point lies on it. As the tool is aligned to x-axis, equation (3) presents the equation of cylindrical surface which is aligned to x-axis, and (y_c, z_c) are the coordinates of the center of this surface and r is the radius of it. Just as the collision is occurring if the line edge (line segment) intersects the polygon (don't only intersect the surface of the polygon), checking process is performed to specify if the intersection point inside the line segment and inside the polygon.

To get an intersection point between line and plane, the coordinates of the equation (1) are substituted in equation (2), then the parameter t could be determined. The coordinates of the intersection point are determined after substituting by the parameter t in the equation (1). Similarly, the intersection points between line and cylindrical surface could be determined. There are two values of the parameter t are gotten by solving this equation because there are two intersection points between line edge and cylindrical surface. After getting the intersection point between line edge and planar surface, the location of intersection point must be specified which is inside or outside this polygon. Figure 4 presents the steps of the test of the intersection point location which inside or outside planar polygon. First, the planar polygon which presented in figure 4(a) is divided into a triangle as shown in figure 4(b), and then the subroutine starts with one of these triangles with area A as shown in figure 4(c). The area of any triangle is determined by half of the cross product of any two vectors of the triangle's sides. There are three triangles which are consisted of intersection point and start point and end point of every edge of the triangle. The areas of these triangles are D, H, and G. As soon as the summation of these areas of these triangles (D+H+G) equal to the area of the original triangle, the intersection point is inside the original triangle (inside the polygon) as shown in figure 4(d). On the other hand, if this summation is larger than the area of the original triangle, the intersection point is outside the original triangle as shown in figure 4(e). This examination is applied to the divided triangles one by one. Finally, if the intersection point is outside all of the divided triangles from the original planar polygon, there isn't intersection between a line and planar polygon. In case of the one of the tool faces is a cylindrical surface polygon, there are two intersection points as mentioned above. If the coordinates of one of the intersection points are located within the coordinate between arc coordinates and line coordinates of the cylindrical polygon, this intersection point inside the cylindrical surface polygon.

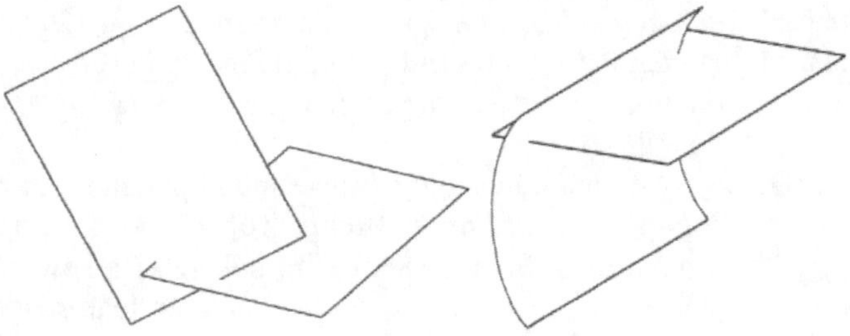

Figure 3. intersection cases between two polygons in 3D

$$(x, y, z) = (x_s, y_s, z_s) + t\,(x_e - x_s, y_e - y_s, z_e - z_s) \tag{1}$$

$$ax + by + cz = d \tag{2}$$

$$(y - y_c)^2 + (z - z_c)^2 = r^2 \tag{3}$$

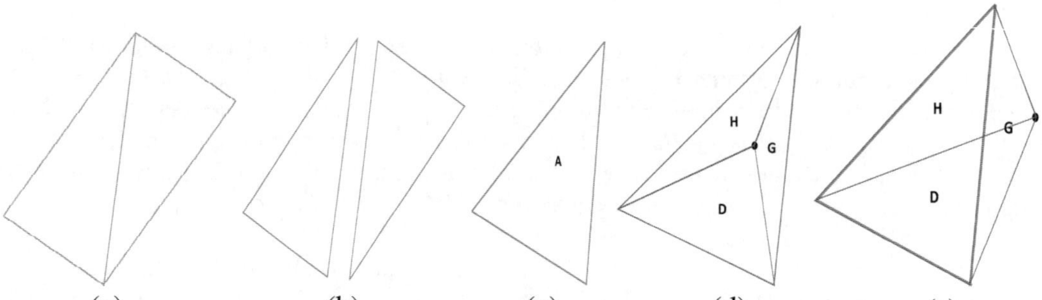

Figure 4. steps of the test of the intersection point location which inside or outside planar polygon: (a) planar polygon, (b) dividing the planar polygon into triangles, (c) one selected triangle with area A, (d) in case of intersection point is inside selected triangle, (e) in case of intersection point is outside selected triangle

4. Conclusions

The proposed intelligent collision detection algorithm in this paper is used to select the V-bending tools automatically by ratifying the feasibility condition of V-bending process planning. This algorithm uses the relations between bend lines to classify any bent part as a 2D or 3D bent part. The collisions between the 2D bent parts and the bending tools could be detected in 2D plane rather than in 3D. Accordingly, this classification of bent parts could reduce the computational loads of the collision detection processes. 2D and 3D collision detection subroutines have been developed in the proposed algorithm to detect the collisions during bending according to the type of bent part. In the future work, suitable optimization technique will link to this collision detection algorithm to search for an optimized bend sequence and bending tools.

5. References

[1] J. R. Duflou, J. Váncza, and R. Aerens, "Computer aided process planning for sheet metal bending: A state of the art," *Comput. Ind.*, vol. **56**, no. 7, pp. 747–771, Sep. 2005.

[2] A. C. Lin and C. Chen, "Sequence Planning and Tool Selection for Bending Processes of 2.5D Sheet Metals," *Adv. Mech. Eng.*, vol. **2014**, 2014.

[3] J. C. Rico, J. M. González, S. Mateos, E. Cuesta, and G. Valiño, "Automatic determination of bending sequences for sheet metal parts with parallel bends," *Int. J. Prod. Res.*, vol. **41**, no. 14, pp. 3273–3299, 2003.

[4] M. Engineering, "Sheet Metal Bending Operation Planning : Using Virtual Node Generation to Improve Search Efficiency," *J. Manuf. Syst.*, vol. **18**, no. 2, pp. 127–139, 1999.

[5] J. R. Duflou, T. H. M. Nguyen, and J.-P. Kruth, "Intelligent tool pre-selection: A contribution to automatic process planning for sheet metal bending," in *Proceedings of the TMCE 2004, April 13-17*, 2004, pp. 671–682.

[6] U. Alva and S. K. Gupta, "Automated design of sheet metal punches for bending multiple parts in a single setup," *Robot. Comput. Integr. Manuf.*, vol. **17**, no. 1–2, pp. 33–47, Feb. 2001.

[7] A. A. Salem, T. F. Abdelmaguid, A. S. Wifi, and A. Elmokadem, "Towards an efficient process planning of the V-bending process: an enhanced automated feature recognition system," *Int. J. Adv. Manuf. Technol.*, 2017.

Comparison and Research of the Mechanical Items of Standards for Controlled Door Closing Devices

X Ren[1], M Wang[1], Z K Li[1] and S M Ou[1]

[1]Foshan Supervision Testing Centre of Quality and Metrology, Foshan 528225, China

willa_job2016@163.com

Abstract. This paper analyzes the standards of controlled door closing devices at home and abroad, compares the standards from test devices and items, and illustrates the backcheck function, delayed action and dead stop degree test in detail. In addition, experiments have been done to verify the influence of different door weights on closing time, opening and closing moment and efficiency. The results show that the door weight has significant influence on mechanical performance of controlled door closing devices.

1. Introduction

Controlled door closing devices are any manually operated door closing mechanism where the energy for closing is generated by the user upon opening the door, and when released returns the door to the closed position, in a controlled manner. The domestic controlled door closing devices are divided into door closer and floor spring, which transmission are pinion and rack, worm gearing respectively. Thus the door closer can open and close only in one direction while the floor spring can open and close in both directions. Besides, floor spring is especially made for heavy doors up to 300 kg. Based on the driving type, domestic controlled door closing devices are divided into hydraulic and electric while these automatic closing devices are called controlled door closing devices in foreign standards.

Controlled door closing devices can not only close the door automatically, but also protect the door frame and door plate, and more importantly, they have became a key executive part in intelligent management of modern buildings, and be widely used in frequent access places, such as commercial and public buildings[1]. With further opening of domestic hardware market, more and more international famous controlled door closing products are coming. At the same time, the quality of many domestic controlled door closing devices have also been improved greatly, and many new types of controlled door closing device have been designed [2-4]. But there still exist differences between the domestic standards and foreign standards, for example, durability is the most important test items in these standards. In domestic standards, the highest requirements are 1000 000 cycles for door closer and 500 000 cycles for floor spring while in America, the requirements are 2000 000 cycles and 1000 000 cycles respectively.

So, comparing the domestic and foreign standards for controlled door closing devices can make a great contribution to the development of domestic products. The current authority international standards are EN 1154-1996(Controlled door closing devices only can be sold in Europe after pass through the tests of this standard)[5], BHMA A156.4-2013 (The highest standard in the world, and controlled door closing devices only can be sold in America and Canada after pass through the tests of this standard) [6] and JIS A 1510-3:2001[7]. Domestic standards for controlled door closing devices consist of QB/T 2697-2013 (Floor spring)[8], QB/T 2698-2013 (Door closer) [9] which are put forward by China national light industry council, and JG/T 268-2010 (Door closer for building) [10]. This paper mainly compares and analyses QB/T 2697-2013, QB/T 2698-2013 with EN 1154-1996 and BHMA A156.4-2013. In these standards, controlled door closing devices are graded into different classes, and the test door weight changes according to the grades of controlled door closing devices. To investigate the influence of different test door weight on the mechanical performance of controlled door closing devices, experiments have been designed and done.

2. Test Apparatus

The position of the test door is a reflection of the mechanical performance of the controlled door closing devices. During the test, the position of the controlled door closing devices should reflect in the test apparatus intuitively. The method to determine the position of test door in QB/T 2697-2013, QB/T 2698-2013 and EN 1154-1996 is placing a floor template illustration on the floor of the test apparatus for angular measurement. The floor plate is just below the test door with the closed position is 0°.

In BHMA A156.4-2013, the methods to determine the position of the test door consist of angel measurement and length measurement. The angel measurement method is the same with EN 1154-1996. The length measurement method is marking lines at 0.5,2,3,4 and 12 inches perpendicular to the door in the closed position, and 30 in. from the pivot centre of the door, see figure 1.

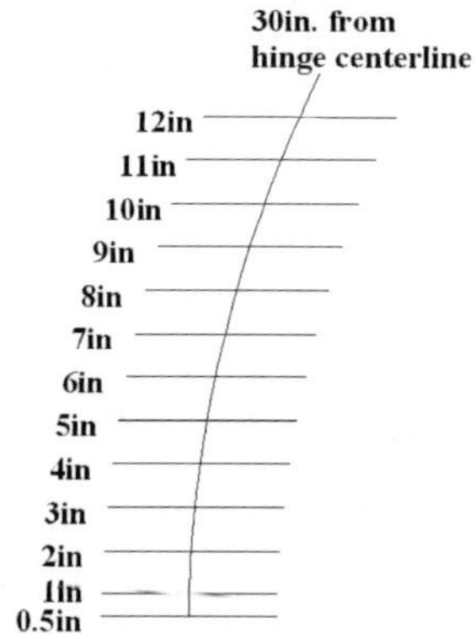

Figure 1. Floor template illustration of BHMA A156.4

Detailed differences of demands for test apparatus and test doors are listed in Table 1.

Table 1. Comparison of the test equipments

	Demands for test apparatus	Demands for test doors	Forcemeter
Floor spring (QB/T 2697-2013) Door closer (QB/T 2698-2013)	Hydraulic: Capable of mechanically opening the door to the 75° and the door can only be closed by door closer; Electric: The door can be opened and closed by the electric door closer without external force.	The size of test door is 900mmx 2000mm. The mass of test door consist of 20 kg, 40 kg, 60 kg, 80 kg, 100 kg, 120 kg, 160 kg, which is determined by the grades of the door closer.	No accuracy requirement.
European standard (EN 1154-1996)	Single action door closers: capable of manual opening to 180° and with automatic actuating means to enable the maximum opening angle of the test requirement; double action door closers: capable of manual opening to 120° and with automatic actuating means to enable the maximum opening angle in alternate directions of the test requirement.	The test door shall be 2000mm high, and of any width between 750 mm and 1200 mm, except that for door closers which do not carry the mass of the test door or act as a door pivot, the test door may be between 400 mm and 2000 mm high. The door mass is the same with QB/T 2697-2013.	The accuracy of the forcemeter is at least 1.5%.
American standard (BHMA A156.4-2013)	Capable of mechanically opening the door to the 90°(±5°) and releasing shall be provided for the cycling test.	Centre of gravity of the weighted door shall be 457 mm from the pivot centre of the door for size 2 or larger closers and 381 mm for size 1 closers. The door mass for size I, II, III, IV, V and VI closers are 23 kg, 36 kg, 45 kg, 57 kg, 70 kg, 82 kg respectively.	The accuracy of the forcemeter is at least 5%.

3. Analysis of Test items

3.1. Backcheck Function

Controlled door closing devices with backcheck function can reduce the impact caused by opening the door, thus avoid damage to the wall or door plate. In QB/T 2697-2013 and QB/T 2698-2013, only hydraulic controlled door closing devices have backcheck function. The test methods and requirements of backcheck function test in different standards have been compared, the differences are as follows:

3.1.1. In BHMA A156.4-2013 and QB/T 2697-2013, QB/T 2698-2013, the backcheck function is tested by observing. At first, set the closing speed as required, then observe the movement of the test door, and record the angel when the speed of the test door have an observable reduction, and the angel when the door stops. In QB/T 2697-2013, QB/T 2698-2013, the test door must slow down obviously after 65°, and stop completely before 90°. In BHMA A156.4-2013, the controlled door closing devices must provide an observable reduction in the door opening speed between 60° and 85° of door opening and the door shall be completely stopped at a maximum of 90°.

3.1.2. In EN 1154-1996, the controlled door closing devices with backcheck function are divided into door closers incorporating an adjustable backcheck function and door closers incorporating a fixed backcheck function.

3.1.2.1. For door closers incorporating an adjustable backcheck function, when the ambient temperature meets the specified requirement， set the closing time regulate(s) according to the requirements, adjust the backcheck function of adjustable backcheck closers to arrest the test door at the 80° open position and continue the cycling test for a total of 1000 000 test cycles. Allow the body surface of door closer to cool to within 2℃ of the original ambient temperature recorded before and verify that the test door is arrested before the 90° open position. The adjustable backcheck shall not be readjusted prior to this test. Upon completion remove the backcheck action and reset the test apparatus as required. Continue cycling for a further 400 000 test cycles.

3.1.2.2. For door closers incorporating a fixed backcheck function, when the ambient temperature meets the specified requirement， set the closing time regulate (s) according to the requirements, the test door shall be arrested by the backcheck at the 90°open position or less. Continue the cycling test for a total of 100 000 test cycles. Allow the body surface of door closer to cool to within 2℃ of the original ambient temperature recorded before and verify that the test door is arrested before the 90° open position. Continue cycling for a further 400 000 test cycles with the test apparatus set to open the test door to an opening angel of between 70° and 75°.

3.2. Delayed action

Controlled door closing devices with delayed action can close the door from the maximum opening position slowly and uniformly, and the speed of closing door can be adjusted steplessly. Controlled door closing devices with delayed action are especially suitable for the frequent in-and-out places, e.g. hospital, access for old man, children and the disabled. In QB/T 2697-2013 and QB/T 2698-2013, only the hydraulic controlled door closing devices have delayed action, and the differences of the test requirements and test methods with EN 1154-1996 and BHMA A156.4-2013 are as follows:

3.2.1. In QB/T 2697-2013 and QB/T 2698-2013, after 5000 test cycles, open the speed regulators completely and close the backcheck valve completely, open the test door to 95° then release. Measure and record the position and delay time of the speed of test door changes obviously from the 90° open position. The delay time from 90° to the end of the delay zone must be at least 10 s, and the end of the delay zone is between 60° and 75°.

3.2.2. In EN 1154-1996, the test method and test parameters are more complex than domestic standards. To test the delayed action, the moment required to move the test door out of the delay zone in a time of 2s to 3s, the door closing time from 90° to the end of the delay zone, and the door closing time from 90° to the end of the delay zone in the last 5 cycles must be measured and recorded. After the tests, the controlled door closing device should satisfied with the moment for the grades-related door closers, the door closing time from 90° to the end of the delay zone after 5000 cycles and 500000 cycles are at least 20 s, the door closing times from 90° to the end of the delay zone in the last 5 cycles of 5500 cycles are all between 10s and 30s.

3.2.3. In BHMA A156.4-2013, at first, fully open both general and latch speed regulating valves and adjust the spring force. Fully close the delayed action regulating valve. Then open the door to 90° and release, the closing time between 90° and 70° shall be a minimum of 20 s. The point at which the door noticeably accelerates shall be a minimum of 60°.

In terms of delayed action, the requirements of EN 1154-1996 are much higher than QB/T 2697-2013, QB/T 2698-2013 and BHMA A156.4-2013, which have the same test method, but the requirements of the latter is higher.

3.3. Hold open

Controlled door closing devices with hold open function can keep the door open into either a preset or chosen angle until manually closed or electrically released, which are used in the places the door need open for a long time, for example, passage of the mall and access for disabled, etc. Hold open function is an inbuilt function in EN 1154-1996, but there is no specific requirement.

3.3.1. In QB/T 2697-2013 and QB/T 2698-2013, only electric controlled door closing devices have hold open function. The test method and requirement are relatively easy, which is giving an opening signal, and checking whether the test door can open for a long time at the setting angle.

3.3.2. In BHMA A156.4-2013, the requirements for hold open are specific. Firstly, the controlled door closing devices with hold open function consist of automatic hold open and selector hold open, and the regulated hold open angel is between 85° and 180°. Besides, in the selector hold open test, the test door can not be opened to the hold open position when the selector turned to off or non-hold-open.

4. Experimental analysis

In these standards, the test door weights are graded into different grades. To work out the influence of different door weights on mechanical performance of controlled door closing devices, experiments have been designed. Combined with existing equipments and products, the grade of tested door closer is 3 and the floor spring is 5. The test door weights are from 30 kg to 110 kg, 10 kg for a grade. The test results are as follows.

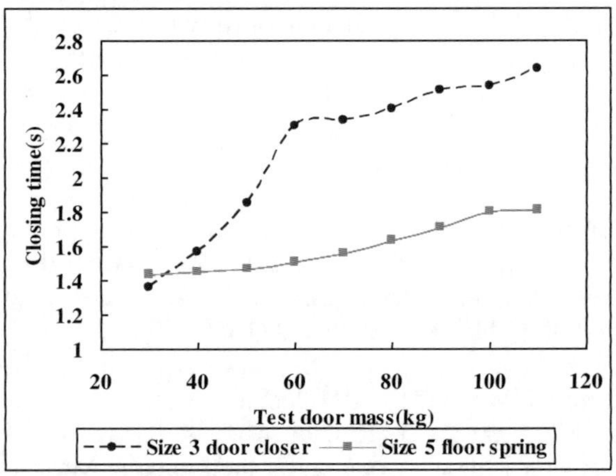

Figure 2. The influence of different door weights on closing time

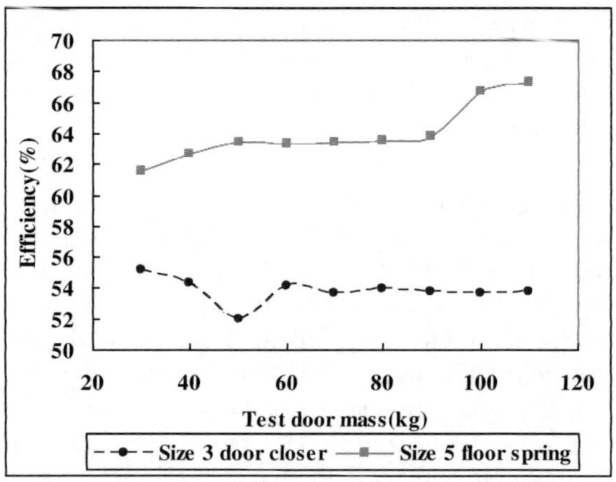

Figure 3. The influence of different door weights on efficiency

From the test results, we know that the test door mass determines the application performance of controlled door closing devices. Relatively speaking, the effect of the door mass on the closing time is bigger than efficiency. And compares the results of the effect of door mass on door closer and floor spring, we can put that the size 5 floor spring is less affected, see figure 2-3. Other mechanical projects, such as overload performance, the result is relatively vague and not so convenient to compare.

5. Conclusion

QB/T 2697-2013 and QB/T 2698-2013 are referred to EN 1154-1996, so the test items are roughly the same. They have significant difference with BHMA A156.4-2013. In general, the test requirements and methods of domestic standards for controlled door closing devices are lower than that of foreign standards.
(1) In the specific test items, the test methods and requirements of EN 1154-1996 and BHMA A156.4-2013 are more detailed and stringent. There are few items that requiring subjective observation.
(2) In terms of application scope, BHMA A156.4-2013 covers the most extensive category, even the new type of door closers which are primarily developing in the country.
(3) On the measuring instruments, EN 1154-1996 and BHMA A156.4-2013 have a clear request for the devices and the requirements are higher.
(4) The requirement of test environment of EN 1154-1996 is the highest, every test item has specific requirement for test environment and speed.
(5) The results of the effect of test door mass on the mechanical performance of controlled door closing devices illustrate that the effect is significant. So before the test, the grade of the controlled door closing devices must be determined and offered by the manufacturers, thus the test house can give accurate test data.
The comparison and analysis between the existing domestic and abroad standards for controlled door closing devices can make the related person understand and implement the standards better. Based on the different requirements of domestic and abroad standards, understand the inadequacies of domestic standards and improve them. Thus improve the test efficiency, and make the research and test meet the demand of current market, thus promote the development of domestic building hardware and improve the export trade volume.

6. References

[1] Ying S S, Ji S M, Xia G Q *et al* 2013 *J.* Design and research of integrated hinges for automatic door closure. *Machine design and research.* **29**(2): 33-35
[2] Ying S S, Ji S M,, Tan D P et al 2012 *P.* An automatic door closing hinge with separated tension and speed adjustment. China: CN102561871A, 2012-07-11
[3] Zhang Z M, Gao Y A, Zhou H 2008 *J.* Development of temperature-sensed automatic door closer based on SMA actuator. *Journal of engineering design.* **15**(6): 431-434
[4] Ruan T, Li B, Diao X L *2013 Conf. on science and technology (Anhui Hefei)* (Brief Discussion on development of the standards for fire door closers) *p* 674-676
[5] EN 1154: 1996. Building hardware–Controlled door closing devices-Requirements and test methods. *S*
[6] ANSI/BHMA A156.4-2013.American national standard for door controls-closers. *S*
[7] JIS A 1510-3: 2001. Test method for door fittings of buildings- Part 3: floor concealed door closers, door closers and hinge closers. *S*
[8] QB/T 2697-2013. Floor Spring. *S*
[9] QB/T 2698-2013. Door Closer. *S*
[10] JG/T 268-2010. Building Door Closer. *S*

2nd International Conference on Design and Manufacturing Engineering (ICDME2017) IOP Publishing
IOP Conf. Series: Materials Science and Engineering **239** (2017) 012004 doi:10.1088/1757-899X/239/1/012004

Design of a New Integrated Structure of the Active Suspension System and Emergency Lane Change Test

ZHAO Jing-bo, LIU Hai-mei, ZHANG Lan-chun, BEI Shao-yi

School of Automotive and Traffic Engineering, Jiangsu University of Technology, Changzhou, China

E-mail: zhaojb1128@jsut.edu.cn

Abstract. An integrated structure of the active suspension system was proposed in order to solve the problem of the individual control of the height of the body or the adjustable damping of the active suspension system of the electric vehicle, which improve the vibration reduction performance of the vehicle. The air bag was used to replace the traditional spiral spring, and the traditional shock absorber was replaced by the damping adjustable shock absorber, and the control module received the body acceleration sensor and the horizontal height sensor signal. The system controlled adjustable damping coefficient of shock absorber through the height of the car body the output of the air pump relay and the height control valve and the output of the electromagnetic valve of the adjustable damping shock absorber, and the emergency lane change test was carried out under different modes of speed of 60km/h. The experimental results indicated that the damping value was greater, average roll angle, yaw angle and average vehicle lateral acceleration were small when vehicle body was in the state of emergency lane change, which verified the feasibility of the integrated control strategy and structure design of the active suspension system. The research has important theoretical research value and engineering application prospect for designing and controlling strategy of vehicle chassis integrated control system.

1. Introduction

Suspension system of vehicle has an important influence on vehicle ride comfort and handling stability. The traditional passive suspension cannot be adjusted due to its damping and stiffness, and it limits the dynamic performance of the vehicle. The active suspension can take into account the ride comfort and handling stability of the vehicle, however, it is not widely used because of its high energy consumption and high cost. Semi-active suspension overcomes the defects of passive suspension, its performance close to active suspension, which has the advantages of high cost performance, low energy consumption. Therefore, semi-active suspension has been a hot research topic in the field of vehicle chassis technology.

In order to give full play to the vibration reduction performance of the suspension system, it is required that the damper has a variable damping coefficient to meet the requirements of damping in different driving conditions [2]. The main way to change the damping coefficient of the shock absorber is to change the viscosity of the damper and adjust the orifice area [3]. Variable damper and electrorheological damper are used as shock absorber for adjusting damping coefficient of damper [4,5]. Mechanical damping adjustable damper, Pneumatic control damping adjustable shock absorber, Electromagnetic valve controlled damping adjustable shock absorber and Motor control damper are also used as Damper for regulating orifice area [6].

The research and development of damping adjustable shock absorber is mainly focused on the performance analysis of adjustable damping shock absorber at home and abroad [9]. Some papers[10,11] have present a series of orifice area adjustable shock absorber and the damping force can be controlled by adjusting the throttle area of the throttle orifice. It is important theoretical research value application and prospects that how the tension damper and the compression stroke presents different damping characteristics [9]. According to the driving conditions, the damping of the damper's stretching and compression stroke is multi-mode switching to meet the more complex conditions of the ride and handling and stability in the same mode.
This paper presents an integrated active suspension system and analyses the structure and working principle of the integrated active suspension system. And dynamic model and control strategy of air spring and shock absorber for emergency lane change test, in order to grasp the technical principle and the basic characteristics of the electronic control suspension system, It also provides a useful reference for the independent research and development of active suspension system.

2. Structure of integrated active suspension system

Active suspension system uses air spring and shock absorber as integral structure of the vibration damping components, the transformation of the damping force is realized by a single valve PDC in the suspension system. The valve is integrated in the shock absorber and is connected with the air spring through a hose, as shown in Figure 1, The pressure of air spring (which is proportional to the load) is used as an adjustable parameter to control the variable orifice on the valve PDC, which affects the flow resistance and influence spring back and the damping force.

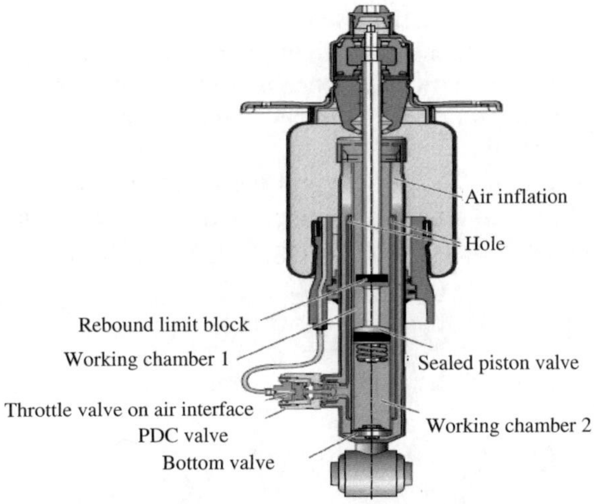

Figure 1. Integrated structure of active suspension system

In order to balance the pressure in the air spring unwilling to change the dynamic pressure (compression and rebound), the air interface of the valve PDC is equipped with a throttle valve. The working process is as follows:
As shown in Figure 2 (a), the elongation process of the air spring when the pressure is smaller. The piston is pulled upward, a part of the oil flow through the piston valve, another part of the oil flow from the working chamber 1 hole to the valve PDC. when the control pressure (air spring pressure) and liquid flow through the valve PDC resistance becomes smaller, so the damping force (damping force) is reduced.
As shown in Figure 2 (b), the elongation of the air spring when the pressure is bigger. The control pressure (air spring pressure) and liquid flow through the PDC valve resistance increased. Most of the liquid (depending on the control pressure) must flow through the piston valve, so the damping force (damping force) increases.
As shown in Figure 2 (c), the compression process when air spring pressure is smaller. The damping force determined by the valve and (to some extent) the liquid which flows through the valve resistance when the piston is pressed downwards, a part of the oil where piston rod extrude flow into the oil

chamber through the bottom valve. Another part of the oil flow to the valve PDC through the hole of the working chamber1.

As shown in Figure 2 (d), the compression process when air spring pressure is bigger. As the control pressure (air spring pressure) and liquid flow through the PDC valve resistance increased. Most of the liquid (depending on the control pressure) must flow through the valve, so the damping force (damping force) is increased.

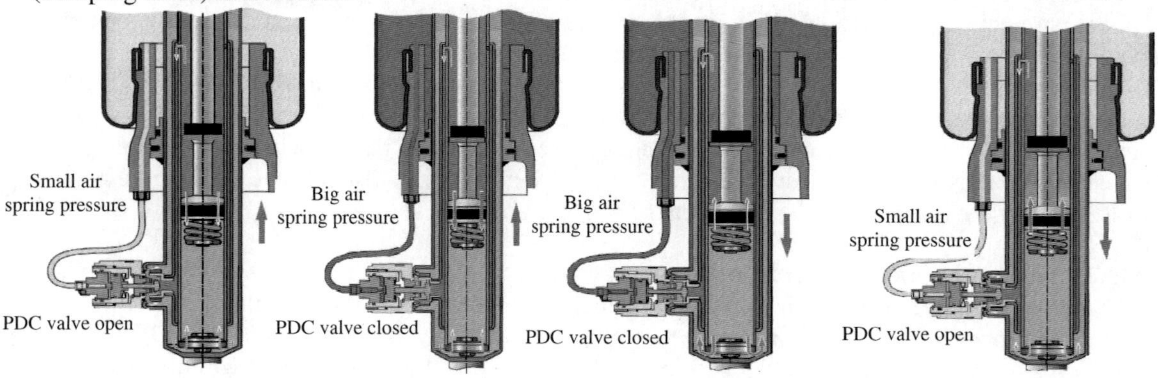

(a) (b) (c) (d)

Figure 2. The compression process under bigger stress

3. Dynamic control method of integrated active suspension system

Air spring is an elastic component of suspension, which has the characteristics of variable rigidity, low vibration frequency and anti-bump. And now it has been more and more widely used. According to the equation of state of the gas, the pressure and volume of the compressed air in the air spring are satisfied when the air spring is working.

The minus sign indicates change trend of pressure and the trend in volume. The pressure increases when the volume is reduced. The absolute value is calculated when the stiffness value is calculated. According to the flow conservation, we get the following:

$$\begin{cases} Q_b = Q_t + Q_c \\ Q_t = Q_1 + Q_{crack} \end{cases} \tag{1}$$

According to the relevant knowledge of the cylindrical annular gap flux, the flow rate expression in the annular gap can be obtained.

$$Q_{crack} = \frac{\pi D h^3}{12\mu l_{crack}} \Delta p_1 + \frac{v_0}{2} \pi D h \tag{2}$$

The spring valve plate is deformed when the damping fluid flows through a one-way valve of the damping control valve, there is a gap between the spring valve and the pipeline, and oil in the gap flow to a one-way valve. We obtain the expression of the flow in a turbulent state based on boundary layer theory.

$$Q_{y1} = 2\pi r[-127.8\nu + (2.5\ln\frac{\delta}{2} + 3 - 2.5\ln\nu + 2.5\ln\sqrt{\frac{\Delta p_s \delta}{2l\rho}}] \tag{3}$$

Q_{y1} is the flow of the damping control valve y_1, r is the outer radius of spring valve plate, that is the distance from the outside to the center of the valve plate after the deformation of the valve plate.δ is the transformation value of valve block. Δp_s is the pressure difference between the two ends of the solenoid valve. ν is the kinematic viscosity of fluid, ρ is the density of damping fluid and l is the slit length.

Domination principle of active suspension system is shown in Figure 3, the control module receives the body acceleration sensor and the horizontal height sensor signal, and the control module analyzes the current road condition according to the body acceleration sensor signal, and the height of the current vehicle is analyzed based on the horizontal height sensor signal. At the same time, the control module obtains the effective information that provided by other control modules through CAN Bus. The control module controls the height of the body by the output of the air pump relay and the height control valve; And by adjusting the output of the electromagnetic damper, the damping coefficient of the damper can be controlled.

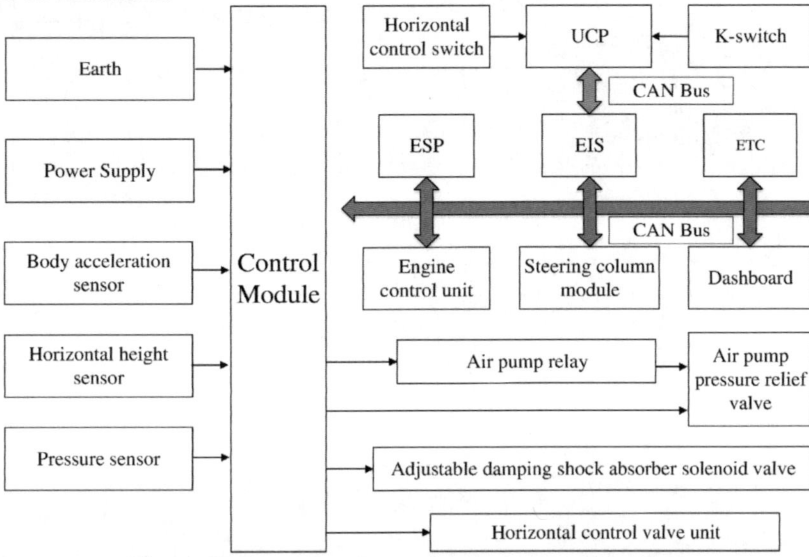

Figure 3. Control strategy of active suspension system

4. Emergency lane change test

The test is a closed loop test for comprehensive determinate the handling stability of "people-vehicle-road", the test is also based on driver's subjective evaluation criteria. The driving dynamic parameters are determined by driving dynamic parameters through the emergency line path, such as steering wheel angle, yaw angle, roll angle, lateral acceleration, etc. Test must be in accordance with the national standard GB /T 6323.3-94 .

The experimental data were collected, such as steering wheel angle, steering wheel torque, roll angle, yaw angle and lateral acceleration data. The experimental data are shown in Figure 4 and Figure 5, and the results of the test data are shown in Table 1 and Table 2.

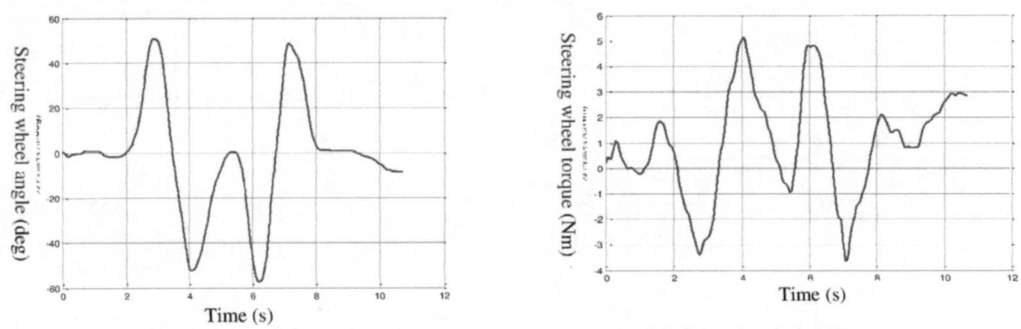

(a) Steering wheel angle (b) Steering wheel torque

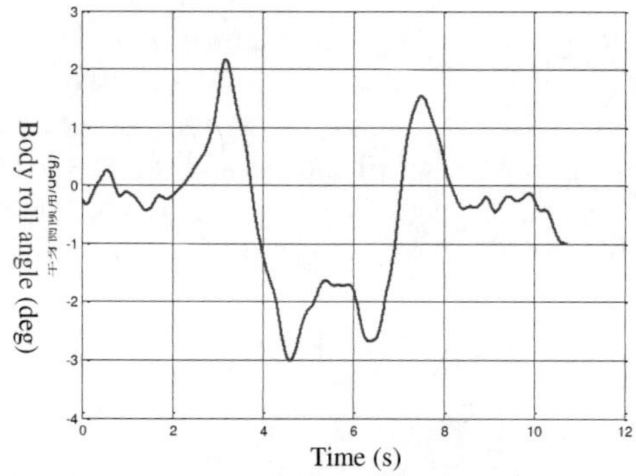

(c) Roll angle

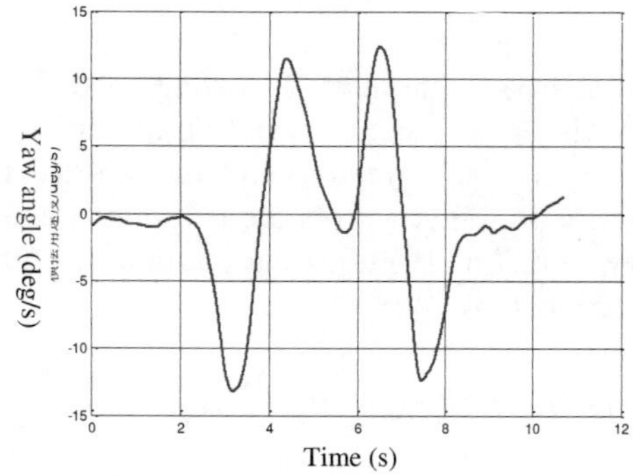

(d) Yaw angle

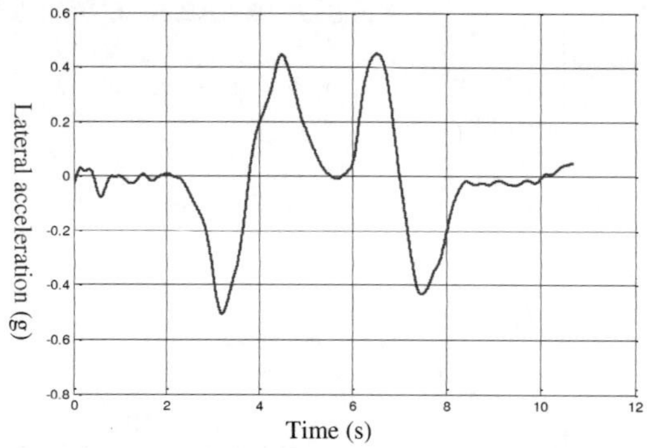

(e) Lateral acceleration

Figure 4. Emergency lane change test results (Hard compression &hard rebound mode, 60km / h)

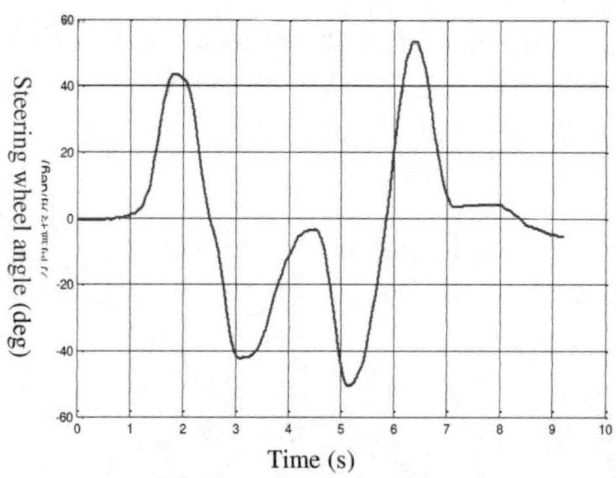

(a) Steering wheel angle

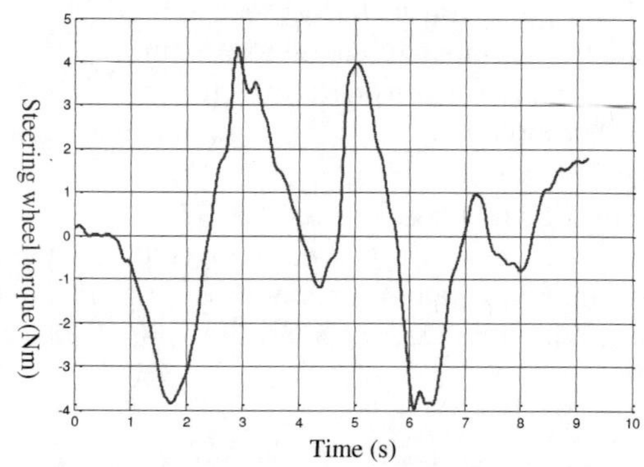

(b) Steering wheel torque

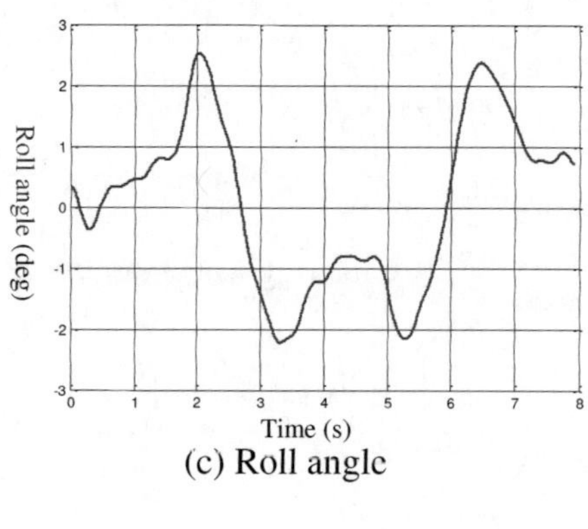

(c) Roll anglc

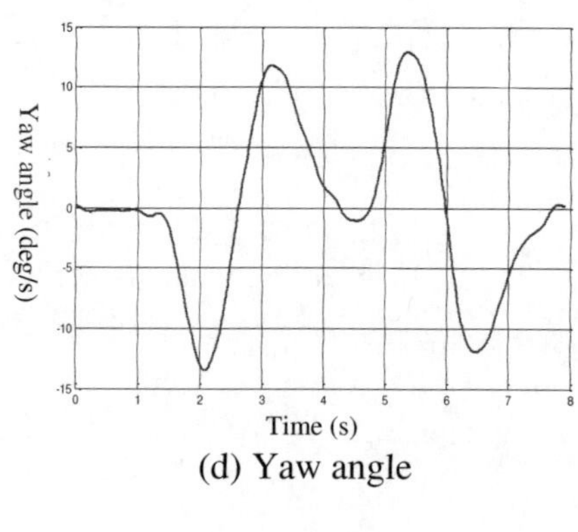

(d) Yaw angle

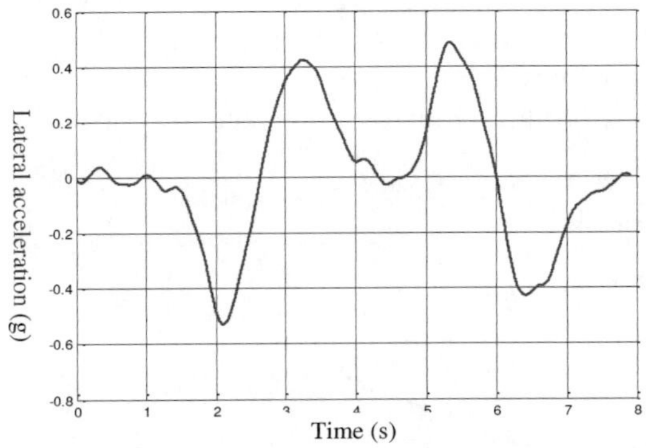

(e) Lateral acceleration

Figure 5. Emergency lane change test results (Hard compression & soft rebound mode, 60km / h)

From Table 1 and Table 2, in general, the average roll angle, the average vehicle yaw angle, the average body acceleration and other indicators with the increase of vehicle speed are increasing under the condition of emergency lane change. This is similar to the situation of the vehicle with the normal passive suspension, which is in accordance with the basic law of the influence of the vehicle speed on the vehicle handling stability.

Table 1. Emergency lane change test indicators (Hard compression &hard rebound mode)

Speed (km/h)	Average steering wheel angle (deg)	Average steering wheel torque (Nm)	Average roll angle (deg)	Average yaw angle (deg/s)	Average lateral acceleration (g)
30	65.5234	3.1140	1.4065	9.7150	0.1895
40	61.1028	3.4355	1.6696	11.2991	0.2741
50	55.6729	3.7262	1.9533	12.1402	0.3705
60	50.0203	4.1959	2.3209	12.1272	0.4444

Table 2. Emergency lane change test indicators (Hard compression & soft rebound mode)

Speed (km/h)	Average steering wheel angle (deg)	Average steering wheel torque (Nm)	Average roll angle (deg)	Average yaw angle (deg/s)	Average lateral acceleration (g)
30	55.7757	2.9701	1.1154	8.5397	0.1619
40	53.8411	3.3084	1.5514	10.1075	0.2561
50	43.0654	3.3869	1.6692	9.5327	0.3024
60	52.7103	4.2290	2.3159	12.5467	0.4658

From Figure 6, Figure 7 and Figure 8, the damping mode of the shock absorber has an effect on the average body angle, the average lateral acceleration and the average yaw angle of the vehicle body in emergency lane change test.

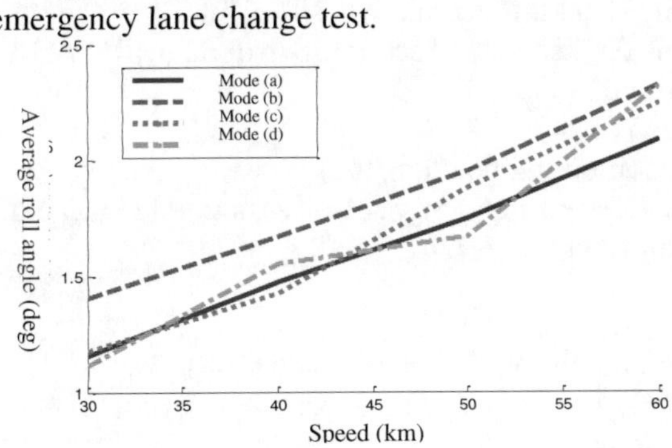

Figure 6. Relationship between speed and the average roll angle

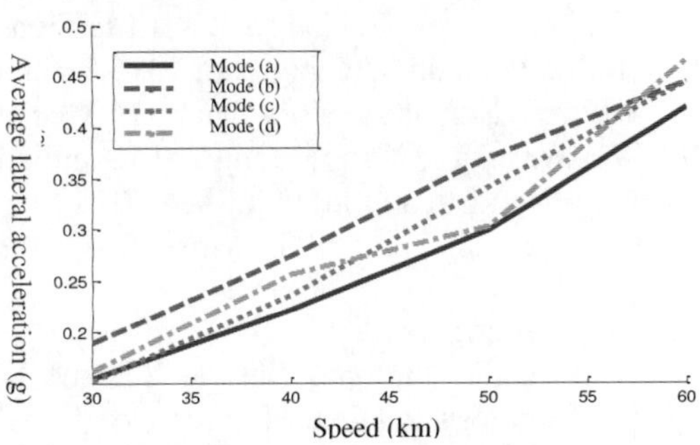

Figure 7. Relationship between speed and the average lateral acceleration

Therefore, the above-mentioned three evaluation indexes are smaller than those of the other three damping modes under the mode of “hard compression &hard rebound”, which is due to the increase of the damping force can effectively inhibit the movement of the piston in the upper and lower chamber. This effectively reducing the tilt of the body, and reduce the average vehicle roll angle and lateral acceleration.

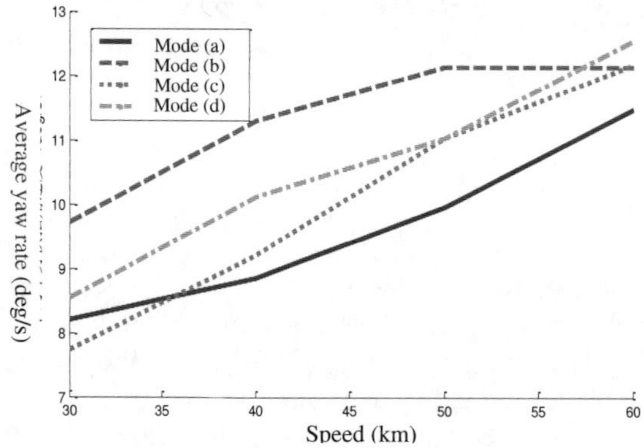

Figure 8. The relationship between speed and the average raw angle

5. Conclusions

The integrated control system of active suspension body height and adjustable damping is replaced by air bag, and damping adjustable shock absorber replace the traditional shock absorber. The height of the body is controlled by the switch between the normal vehicle height mode, the body lift mode and the body down mode, and the adjustable damping control can be used to switch between mode of soft compression & soft rebound, mode of hard compression &hard rebound, mode of soft compression & hard rebound, mode of hard compression & soft rebound. Active suspension system of vehicle body height and adjustable damping integrated control strategy and integrated structure design is reasonable and feasible through the different modes for vehicle speed off hunting test at the speed of 60 km/h road test.

6. References

[1] TUNG S L and JUANG Y T 2011 Expert Systems with Applications vol 38 p 14044
[2] Gysen B L J and Janssen J L G 2011 IEEE Transactions on Vehicular Technology vol 60 p 1384
[3] Kinagi G V and Pitchuka S P 2012 Commercial Vehicle Engineering Congress vol 2 p 111
[4] Zehsaz M and Sadeghi M H 2011 Journal of Terramechanics vol 48 p 439
[5] Wang Ruochen and Zhang Xiaoliang 2012 Transactions of the Chinese Society of Agricultural Engineering vol 28 p 75
[6] Jinshan Lin 2012 Energy Procedia vol 7 p 534
[7] Rongrong Wang and Hui Jing 2015 Journal of the Franklin Institute vol 352 p 468
[8] Li Zhongjie and Lei Zuo 2013 IEEE Transactions on Vehicular Technology vol 62 p 1065
[9] Huihui Pan and Weichao Sun 2015 Mechatronics vol 30 p 363
[10] Weichao Sun and Huihui Pan 2014 Mechatronics vol 24 p 318
[11] Wei Wang and Yuling Song 2015 Applied Soft Computing vol 28 p 167
[12] Jeng-Wen Lin and Pu Fun Shen 2015 Mechanical Systems and Signal Processing vol 62-63 p 63
[13] I. Maciejewski and S. Glowinski 2014 Mechatronics vol 24 p 1242

Acknowledgments
Funding from the National Natural Science Foundation of China (Grant No: 61503163), the "333 project" of Jiangsu Province (Grant No: BRA2016440) and the six talent peaks project in Jiangsu Province (Grant No: ZBZZ-024) are gratefully acknowledged.

IOP Conf. Series: Materials Science and Engineering **239** (2017) 012005 doi:10.1088/1757-899X/239/1/012005

Fixture Locating Precision Pre-Analysis for One-plane-dual-pins fixture System

Liu Wei-dong*, Guo Jinhui

College of Automatic Science and Engineering, South China University of Technology, Guangzhou, 510640

aulwd@scut.edu.cn

Abstract. Type two assembly is mainly fixed by additional fixture but not its own features, and the one-plane-dual-pins fixture system is commonly used for type two assembly assurance; hence, the locating pre-analysis for one-plane-dual-pins fixture system is introduced. The locating variation in one-plane-dual-pins fixture system leads to the gesture displacement of the workpiece in space. Variation source that affects the fixture locating precision mainly includes the fixture locating error and the fabrication error of the fitting holes. While the fixture error and fabrication error of the fitting holes are confirmed, the calculation for the fixture locating error of assembled workpiece is introduced. The fixture locating error simulation is further researched, and one example for fixture locating error propagation is illustrated.

1. Introduction

There are generally two types of assembly during the assembly process: 1) the assembly between two parts or components by fitting operation, where the latter part or component is located according to the first part basis without specific fixture; 2) the assembly between two parts or components by riveting or welding, and each parts or components is located by specific fixture individually. The typical specific fixture layout for pre-locating the assembled workpiece is one-plane-dual-pins layout system. One-plane is usually the large surface for fixing the part, and one of the dual-pins is cylindrical pin, the other pin is the prism pin.

Pre-locating analysis of one-plane-dual-pins layout system is significant for ensuring the assembly accuracy of type two assembly. Wu et al. [1] introduced models for different types of fixture layout system based on the mechanism mapping, and it provides the foundation model for the analysis of locating accuracy. Luo et al. [2] studied the locating problem of any space surface, and analyzed the positioning problem by distance function. Liu et al [3] established a positioning mapping model between the element fixture and the workpiece to minimize the locating error. Tsai et al. [4] proposed a matrix analysis method for the positioning design of prismatic parts.

For actual practice application, the locating accuracy problem of one-plane-dual-pins layout system is discussed in this paper. Size error of the locating pins and holes are considered, and pre-locating analysis of one-plane-dual-pins layout system is introduced.

2. Error source of one-plane-dual-pins fixture system

For type two assembly, the one-plane-dual-pins fixture system is shown in Fig. 1. The locating basis is $Oxyz$, the cylindrical pin is P_1, and the prism pin is P_2. The workpiece can translate along the positioning groove that is α angle with x-axis.

The factors that affecting the locating error of one-plane-dual-pins fixture system include:

(1) The locating error of P1 and P_2: ΔX_{P_1}, ΔZ_{P_1}, ΔX_{P_2} and ΔZ_{P_2}.

(2) The locating error of the fitting holes of workpiece. The hole fitting with P_1 is Y_1, and the hole fitting with P_2 is Y_2; the locating error of the fitting holes of workpiece are: ΔX_{Y_1}, ΔZ_{Y_1}, ΔX_{Y_2} and ΔZ_{Y_2}.

(3) While the assembled workpieces are more than three, the repeated locating error of the part is included.

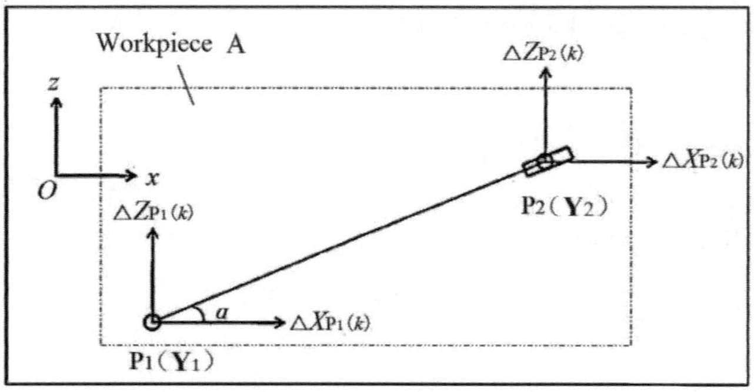

Figure 1. fixture locating of the part in x-z plane

3 Fixture error and locating hole error

3.1 Assembly displacement caused by fixture error

Suppose that the fixture error in process k caused by P_1 and P_2 can be recorded as:

$$\Delta P(k) = [\Delta X_{P_1}(k), \Delta Z_{P_1}(k), \Delta X_{P_2}(k), \Delta Z_{P_2}(k)]^{T} \tag{1}$$

According to reference [5], the locating error of workpiece A caused by the fixture error of P_1 and P_2 in process k is:

$$\boldsymbol{X}_{A}^{L}(k) = \boldsymbol{M}_{AP_1}\boldsymbol{Q}_{P_1P_2}(k)\Delta P(k) \tag{2}$$

Where $\boldsymbol{Q}_{P_1P_2}(k)$ is

$$\boldsymbol{Q}_{P_1P_2}(k) = \begin{bmatrix} 1 & 0 & 0 & 0 \\ 0 & 1 & 0 & 0 \\ \dfrac{\sin\alpha}{L^F} & \dfrac{-\cos\alpha}{L^F} & \dfrac{-\sin\alpha}{L^F} & \dfrac{\cos\alpha}{L^F} \end{bmatrix} \tag{3}$$

$$\boldsymbol{M}_{AP_1} = \begin{bmatrix} 1 & 0 & -L_Z(A,P_1) \\ 0 & 1 & -L_X(A,P_1) \\ 0 & 0 & 1 \end{bmatrix} \tag{4}$$

Where L^F is the distance between P_1 and P_2; $L_Z(A,P_1) = Z_{P_1} - Z_A$, $L_X(A,P_1) = X_{P_1} - X_A$, which are the distances along x and z directions between the centroid point of workpiece and P_1.

3.2 Assembly displacement caused by error of the fitting hole and groove

The locating accuracy also affected by the fabricating error of the fitting hole Y_1 and the fitting groove Y_2. As shown in Fig. 2, suppose that Y_1,Y_2 are the ideal positions of fitting hole and groove, and Y_1', Y_2' are actual positions of fitting hole and groove, the rotation offset of the workpiece is ε.

Suppose that the fabricating error of the fitting hole and groove in process k caused by Y_1 and Y_2 can be recorded as:

$$\Delta Y = [\Delta X_{Y_1}(k), \Delta Z_{Y_1}(k), \Delta X_{Y_2}(k), \Delta Z_{Y_2}(k), \beta(k)]^{T} \tag{5}$$

Where $\Delta X_{Y_1}(k)$, $\Delta Z_{Y_1}(k)$ are translation displacement along the x-axis and z-axis direction caused by Y_1, and $\Delta X_{Y_2}(k)$, $\Delta Z_{Y_2}(k)$ are translation displacement along the x-axis and z-axis direction caused by Y_2, and $\beta(k)$ is the rotation displacement of the fitting groove.

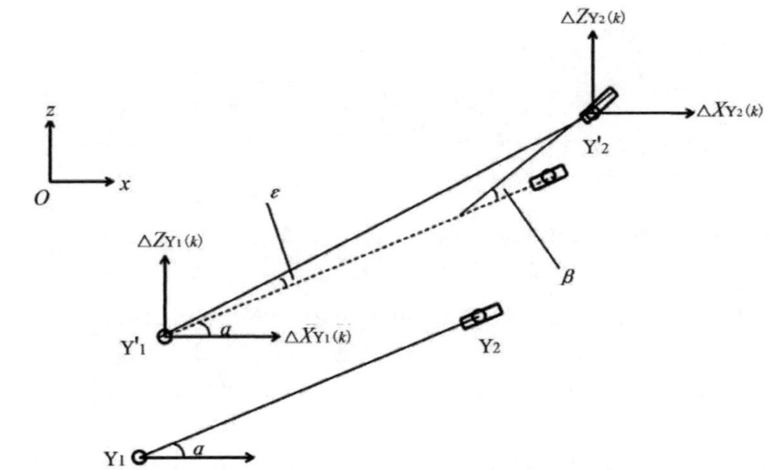

Figure 2. locating accuracy affected by the fabricating error of the fitting hole and groove

The locating error of workpiece A caused by the fabricating error of the fitting groove Y_2 in process k is shown in Fig.3. The ideal position of the prism pin is P_2 and the actual position of the prism pin is P_2'. The rotation offset of the workpiece A is ε, and the lengths of Y_1' P_2 and Y_1' P_2'are r.

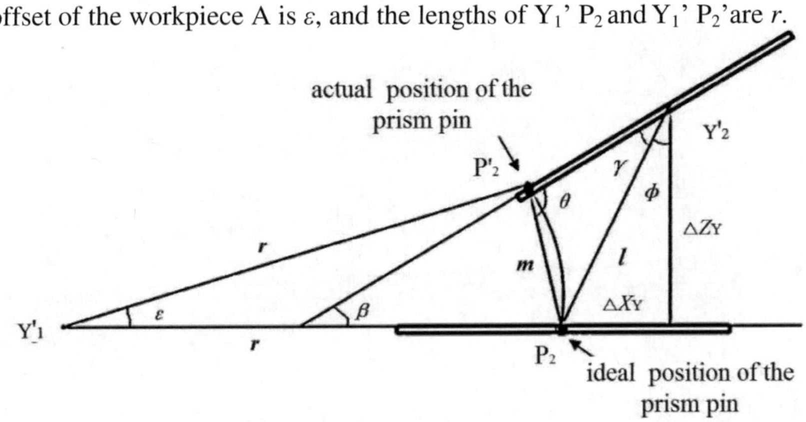

Figure 3. locating error of workpiece A caused by the fabricating error of fitting groove Y_2

As shown in Fig. 3, the displacement of Y_2' referring to Y_1' in the x-axis direction is $\Delta X_Y(k) = \Delta X_{Y_1}(k) - \Delta X_{Y_2}(k)$, in the z-axis direction is $\Delta Z_Y(k) = \Delta Z_{Y_1}(k) - \Delta Z_{Y_2}(k)$. In addition, the following six equations exist:

$$l = \sqrt{(\Delta X_Y)^2 + (\Delta Z_Y)^2} \tag{6}$$

$$\alpha = \arctan \frac{\Delta X_Y}{\Delta Z_Y} \tag{7}$$

$$\beta + \gamma + \alpha = \frac{\pi}{2} \tag{8}$$

$$\frac{m}{\sin\gamma} = \frac{l}{\sin\theta} \tag{9}$$

$$\theta = \beta + \frac{\pi - \varepsilon}{2} \tag{10}$$

$$m \approx 2r\sin\frac{\varepsilon}{2} \tag{11}$$

Solving the equations, the rotation offset of the workpiece is ε:

$$\varepsilon = \beta + \arcsin\left[\frac{\sin(\frac{\pi}{2} - \beta - \arctan\frac{\Delta X_Y}{\Delta Z_Y})\sqrt{(\Delta X_Y)^2 + (\Delta Z_Y)^2}}{r} - \sin\beta\right] \tag{12}$$

The locating error of workpiece A caused by the fitting hole Y_1 in process k is:

$$\boldsymbol{X}_{Y_1}^{L2}(k) = [\Delta X_{Y_1}, \Delta Z_{Y_1}, \varepsilon]^T \tag{13}$$

The locating error of workpiece A caused by the fitting groove Y_2 in process k is:

$$\boldsymbol{X}_{A}^{L2}(k) = \boldsymbol{M}_{AY_1}[-\Delta X_{Y_1}\ -\Delta Z_{Y_1}\ -\varepsilon]^T \tag{14}$$

According to Eq. (12), the ΔX_Y-ε, ΔZ_Y-εsimulation are particularly introduced with β = 5°, 15°, 30°, 60° and 90°.

(1) If ΔZ_Y = 0.1cm, r = 50cm, the ΔX_Y-ε function curve is shown in Fig.4. As shown in Fig. 4, the ΔX_Y-ε function curve is distributed in the first and third quadrant. The larger value of ΔX_Y, can result in the smaller value of ε; while ΔX_Y <-0.0001, the larger value of β, can result in the larger value of ε; while ΔX_Y > -0.0001, the larger the value of β, can result in the smaller value of ε.

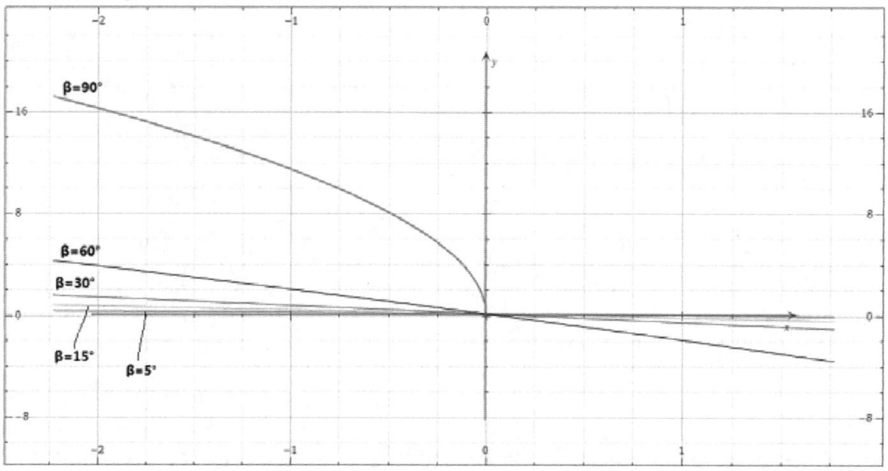

Figure 4. ΔX_Y - ε function curve

(2) If ΔX_Y = 0.1cm, r = 50cm, the ΔZ_Y-ε function curve is shown in Fig.5. As shown in Fig. 5, the ΔZ_Y-ε function curve is distributed in the first and second quadrant. In the second quadrant, the larger value of ΔX_Y, can result in the smaller value of ε; while ΔX_Y <-3.161, the larger value of β, can result in the smaller value of ε; while ΔX_Y > -3.161, the larger the value of β, can result in the larger value of ε. In the first quadrant, the larger value of ΔX_Y, can result in the larger value of ε; the larger value of β, can result in the larger value of ε.

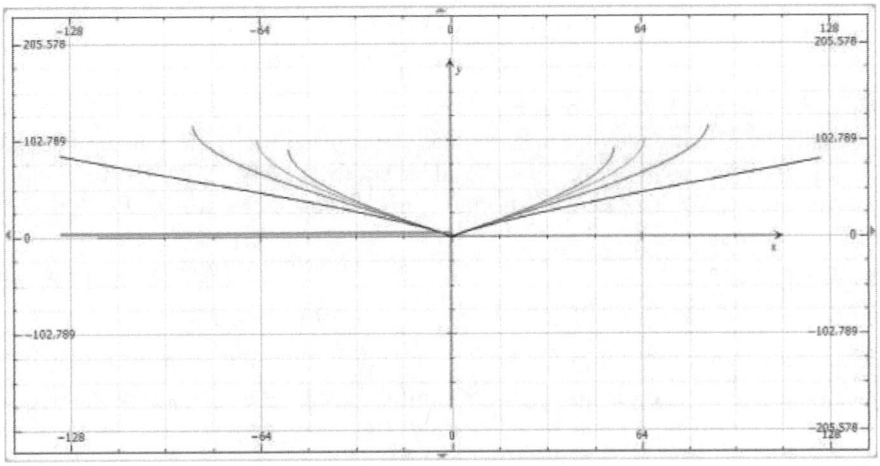

Figure 5. ΔZ_Y - ε function curve

4. Case Study

An assembly with three workpieces is illustrated as an example to verify the calculation of locating accuracy. Fig. 6 is the individual station for this assembly; the dotted line shows the nominal position of parts, and the solid line shows the actual position of parts.

Suppose that the two positioning pins of part one are P_1, P_2; respectively, and the positioning groove are Y_1,Y_2. The two positioning pins of part two are P_3, P_4, and the positioning groove are Y_3,Y_4. Assuming that (1) the fixture error of part one is $\Delta \boldsymbol{P}_{11} = [1,1,1,2]$, the angle α_1 referring to the x-axis is 60°, and the distance between the two pins is L^{F1} = 500mm. (2) the locating error of part one is $\Delta \boldsymbol{Y}_1 = [1,2,1,2, \frac{\pi}{180}]$, the distance between the fitting hole and fitting groove is R_1 = 500mm. (3) the fixture error of part two is $\Delta \boldsymbol{P}_{12} = [1,1,2,1]$, the angle α_2 referring to the x-axis is 60°, and the distance between the two pins is L^{F2} = 500mm. (4) the locating error of part two is $\Delta \boldsymbol{Y}_2 = [2,1,2,1, \frac{\pi}{180}]$, the distance between the fitting hole and fitting groove is R_2 = 500mm.

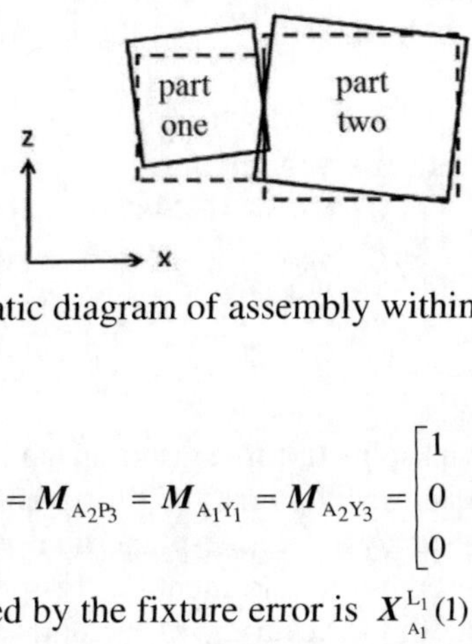

Figure 6. Schematic diagram of assembly within individual station

Within the assembly, suppose that:

$$\boldsymbol{M}_{A_1P_1} = \boldsymbol{M}_{A_2P_3} = \boldsymbol{M}_{A_1Y_1} = \boldsymbol{M}_{A_2Y_3} = \begin{bmatrix} 1 & 0 & 1 \\ 0 & 1 & -1 \\ 0 & 0 & 1 \end{bmatrix}$$

The locating error of part one caused by the fixture error is $\boldsymbol{X}_{A_1}^{L_1}(1) = [1.00002, 0.99998, 0.00002]^T$, and the locating error of part two caused by the fixture error is $\boldsymbol{X}_{A2}^{L_1}(1) = [0.99997, 1.00003, -0.00003]^T$. The locating error of part one caused by the fabricating error of fitting hole and groove is $\boldsymbol{X}_{A1}^{L_2}(1) = [-1.00197, -1.99803, -0.00197]^T$, and the locating error of part two caused by the fabricating error of fitting hole and groove is $\boldsymbol{X}_{A2}^{L_2}(1) = [-2.00197, -0.99803, -0.00197]^T$. According to equation reference [5], the variable flow of station one is:

$$\boldsymbol{X}(1)=\begin{bmatrix}-0.00195 & -0.99805 & -0.00199\\ -1.002 & 0.002 & 0.002\\ 0 & 0 & 1\end{bmatrix}$$

Similarly, for individual station two, suppose that the subassembly after assembly station one is recorded as {1,2}; the two positioning pins of subassembly {1,2} are P_1, P_2; respectively, and the positioning groove are Y_1,Y_2. The two positioning pins of part three are P_5, P_6, and the positioning groove are Y_5,Y_6. Assuming that:

(1) the repeated locating error of part one is $\boldsymbol{X}_{A1}^{L_2}(2)=[0.00195,\ 0.99805,\ 0.00199]^{T}$, the repeated locating error of part two is $\boldsymbol{X}_{A2}^{L_2}(2)=[0.2185,\ 0.00305,\ 0.00199]^{T}$;

(2) the fixture error of subassembly {1,2} is $\Delta\boldsymbol{P}_{21}=[1,1,1,2]$, the angle α_1 referring to the x-axis is 60°, and the distance between the two pins is L^{F1} = 500mm; the locating error of part one is $\Delta\boldsymbol{Y}_1=[1,2,1,2,\frac{\pi}{180}]$, the distance between the fitting hole and fitting groove is R_1 = 500mm.

(3) the fixture error of part three is $\Delta\boldsymbol{P}_{23}=[0.8,0.3,0.6,0.4]$, the angle α_2 referring to the x-axis is 60°, and the distance between the two pins is L^{F3} = 500mm. (4) the locating error of part three is $\Delta\boldsymbol{Y}_3=[0.9,1.1,1,2,\frac{\pi}{180}]$, the distance between the fitting hole and fitting groove is R_3 = 500mm.

Within the assembly, suppose that:

$$\boldsymbol{M}_{A_3P_5}=\boldsymbol{M}_{A_3Y_5}=\begin{bmatrix}1 & 0 & 1\\ 0 & 1 & -1\\ 0 & 0 & 1\end{bmatrix}$$

The locating error of part one caused by the fixture error is $\boldsymbol{X}_{A_1}^{L_1}(2)=[1.00002,\ 0.49998,\ 0.00002]^{T}$; the locating error of part two caused by the fixture error is $\boldsymbol{X}_{A_2}^{L_1}(2)=[1.00028,\ 0.48823,\ 0.00002]^{T}$, and the locating error of part two caused by the fixture error is $\boldsymbol{X}_{A_3}^{L_1}(2)=[0.80001,\ 0.29999,\ 0.00001]^{T}$. The locating error of part one caused by the fabricating error of fitting hole and groove is $\boldsymbol{X}_{A_1}^{L_2}(2)=[-1.00197,\ -1.99803,\ -0.00197]^{T}$, the locating error of part two caused by the fabricating error of fitting hole and groove is $\boldsymbol{X}_{A_2}^{L_2}(2)=[-1.02166,\ -1.01504,\ -0.00197]^{T}$, and the locating error of part three caused by the fabricating error of fitting hole and groove is $\boldsymbol{X}_{A_3}^{L_2}(2)=[-0.93291,\ -1.06709,\ -0.03291]^{T}$ According to equation reference [5], the variable flow of station two is:

$$\boldsymbol{X}(2)=\begin{bmatrix}-0.00195 & -0.99805 & -0.00195\\ -0.80488 & -0.50376 & 0.00125\\ -0.1329 & -0.7671 & -0.0039\end{bmatrix}$$

5. Summary

The locating variation in one-plane-dual-pins fixture system is mainly caused by the fixture error of the positioning pin, and is caused by the fabricating error of fitting hole and groove of each part. For type two assembly, locating accuracy pre-analysis of one-plane-dual-pins fixture system is important for pre-analysis of the final functional assembly requirement. In this paper, the locating and direction error of two-pin positioning part are analyzed; the accuracy of the parts with fabricating error of fitting hole and groove are further calculated. Considering the assembly with three parts, an example is introduced for calculation verification.

6. References

[1] WU Yuguang, ZHANG Genyuan, LI Chunguang. Approach to automated modeling of fixture location errors analysis [J]. Journal of Mechanical Engineering, 2012, 48(5): 172-179 (in Chinese)

[2] LUO Chen, ZHU Limin, DING Han, et al. Statistical characteristic and relative error among fixturing analysis models [J]. Journal of Mechanical Engineering, 2013,45(9): 110-115 (in Chinese) [3] LIU Wenlin, XIONG Caihua. Model of locating error for fixtures [J]. Journal of Huazhong University of Science and Technology: Nature Science, 2003, 31 (7): 72-74 (in Chinese).

[3] LIU Wenlin, XIONG Caihua. Model of locating error for fixtures [J]. Journal of Huazhong University of Science and Technology: Nature Science, 2003, 31 (7): 72-74 (in Chinese).

[4] TSAI Y.H , LU S.S. Localization accuracy matrix analysis for localization accuracy consideration in 2D precision fixture system [C] //Proceedings of 2007 Mediterranean Conference on Control and Automation, doi: 10. 1109/ MED. 2007. 4433966.

[5] TIAN Zhaoqing, LAI Xinmin, LIN Zhongqin. State space model of variaitions stream propagation in multistation assembly processes of sheet metal [J]. Journal of Mechanical Engineering, 2007, 43(2):202-209 (in Chinese)

7. Acknowledgement

This research was financially supported by the National Natural Science Foundation of China (Grant No. 51305140), the Science and Technology Planning Project of Guangdong Province (Grant No. 2014A040401019, 2015A030401029).

IOP Conf. Series: Materials Science and Engineering **239** (2017) 012006 doi:10.1088/1757-899X/239/1/012006

Human Factors and Their Effects on Human-Centred Assembly Systems – A Literature Review-Based Study

Q Wang and M I Abubakar

School of Engineering, University of Portsmouth, Portsmouth PO1 3DJ, UK

qian.wang@port.ac.uk

Abstract. If a product has more than one component, then it must be assembled. Assembly of products relies on assembly systems or lines in which assembly of each product is often carried out manually by human workers following assembly sequences in various forms. It is widely understood that efficiency of assembling a product by reducing assembly times (therefore costs) is vital particularly for small and medium-sized manufacturing companies to survive in an increasingly competitive market. Ideally, it is helpful for pre-determining efficiency or productivity of a human-centred assembly system at the early design stage. To date, most research on performance of an assembly system using modelling simulation methods is focused on its "operational functions". The term used in a narrow sense always indicates the performance of the "operational system", which does not incorporate the effect of human factors that may also affect the system performance. This paper presents a research outcome of findings through a literature review-based study by identifying possible human factors that mostly affect the performance on human-centred manufacturing systems as part of the research project incorporating parameters of human factors into a DES (discrete event simulation) tool.

1. Introduction

Manual assembly is, by definition, carried out by human workers and therefore this type of manufacturing system is human-centred as its performance largely depends on humans (or human performances) rather than machines. There are some studies in a view of socio-technical or psychological sciences to evaluate the effect of human factors relating to the design of manufacturing systems. These human factors may include cognitive elements (such as experience, IQ level and so on) and physical elements (age, gender, dexterity and so on). However, the outcomes of these studies are basically descriptive in a form of language that manufacturing engineers often find difficult to understand. Modelling techniques are used to consider such human issues as ergonomics to improve the environment for workers to perform at a workplace or in a production area with better conditions. Although this improvement may be useful to achieve a better performance of individual workers, it cannot provide an answer of human performance that may impact the overall system performance of such as a human-centred assembly line. Thus, a consideration of human performance often becomes a situation of uncertainty in designing a manufacturing system when human workers or operators are involved in production. In practice, this issue is often neglected by manufacturing system designers, and an adjustment of it in the actual system has to be made based on personal experiences and judgement from production managers. In other aspect, most modelling simulation practices in the production research literature focused on conventional systems in which operators are tied to specific tasks or stations. Moreover, current modelling simulation tools in the market for manufacturing systems evaluation do not provide facilities that allow system designers to combine human attributes (or human performance) into an investigation of the overall system performance. This is because, for example, in a discrete event simulation (DES) model, workers are defined and treated as one of resources the same as parts, machines, conveyors and so on. The application of DES models is therefore restricted to predicting only such variables as the required number of workers, their shift patterns and routes. As a result of this, the accuracy of established DES models can be adversely affected without considering the effect of human performance [1, 2]. This paper presents a research work by investigating human factors and their mutual interactions relating to human performance using findings of a literature review-based study. The study was aimed at identifying and selecting the possible and most influential human factors against a number of key performance measures of human-centred manufacturing systems. The result shows that this approach may provide a useful basis for exploring the existing DES tools by incorporating considerations of human factors that may affect a human-centred manufacturing system into the DES models.

2. Human factors and effects of human performance

Human factors may be referred to such human attributes as their physical and knowledgeable abilities, physiological states and psychological traits. Giniger et al examined the effects of age and experience in the relevance to worker performance, absenteeism, safety and turnover, and suggested that it is the job experience rather than the age which determines the work performance [3]. Avolio et al confirmed this outcome using a polynomial regression analysis to predict the work performance in connection with age and experience [4]. By contrast, McEvoy et al presented a study auguring that there is no clear relationship between age and job performance [5]. Zwick et al observed that the average muscle strength (aerobic capacity) of a worker decreases by roughly 10% per decade over ages from 20 to 60, 15% from 60 to 80 and 30% after 80, i.e., around 1% per year [6]. Adam et al investigated the relationship between general cognitive ability and reaction time [7]. Berg et al stated that the reaction time can be affected by distraction and mental fatigue [8]. Llmarinen observed that age negatively affects general cognitive abilities and positively affects experience; however, experience positively affects cognitive skills and directly affects job performance [9]. Kenny et al found a decrease in the aerobic and musculoskeletal capacity that leads to an average drop of 20% in the physical work capacity of workers aged from 40 to 60 [10]. Figure 1 shows the effects of human factors (age, experience and cognitive ability) on human performance against a number of performance measures (throughput, cycle time, in process inventory and cost) on a human-centered manufacturing system.

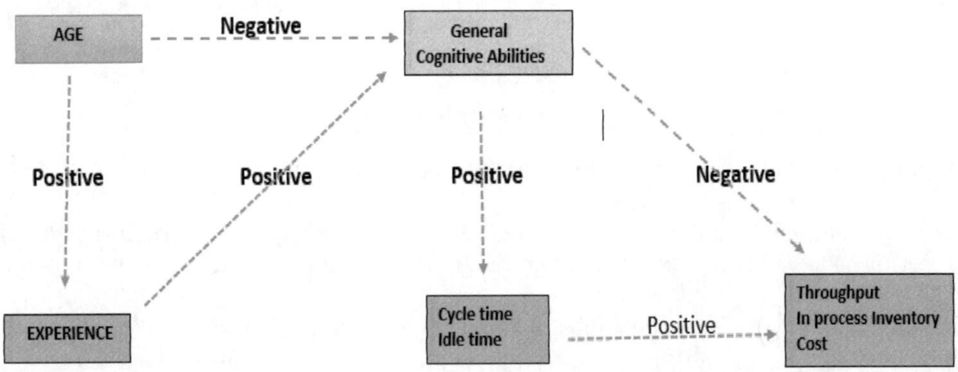

Figure 1. Effects of human factors on human performance against performance measures.

Learning and forgetting are both natural phenomena that directly affect the performance of individual workers. Two major factors of assembly processes, which mostly affect worker performances, are product variety and task complexity. As the product variety increases, it becomes more difficult to learn and easier to forget; higher complexity of selected manufacturing tasks may degrade worker proficiency for product assembly. The learning curve, also known as experience curve, describes the improved efficiency or performance obtained from repeating an operation of a specific task by a worker; i.e. the time required to perform a task declines at a decreasing rate as experience with the task increases. Figure 2 illustrates a typical behaviour of a learning curve in an assembly process that is carried out by a fitter [1]. Performance improvement is often measured in terms of cycle time reduction, which is the result of the learning process. The learning process can be modelled using a DES tool to describe the physical elements and their logical interrelationship with specific rules, which interacts with external MS Excel worksheets to capture individual human workers' behaviours or human factor variables (such as worker experience, age and dexterity) in order to generate the variation of human performance by individual workers. This can be done by setting a value of each human variable to weight such as a worker's performance, which reflects to their working speed as an example. For instance, assuming that a walking-worker has an initial processing time of 150 s per unit (shown in Figure 2) at the beginning of a learning process, we name this amount of time as assembly learning limit (ALL). This value of processing time then declines during a learning process for this selected walking-worker until it reaches a steady state after producing a certain number of units of this product by repetitively performing the same amount of assembly tasks for making a unit of this product. This method enables system designers to evaluate a walking-worker's varying capability in terms of working speed per unit through a learning curve to meet a minimum proficient level of skills acquired from trainings.

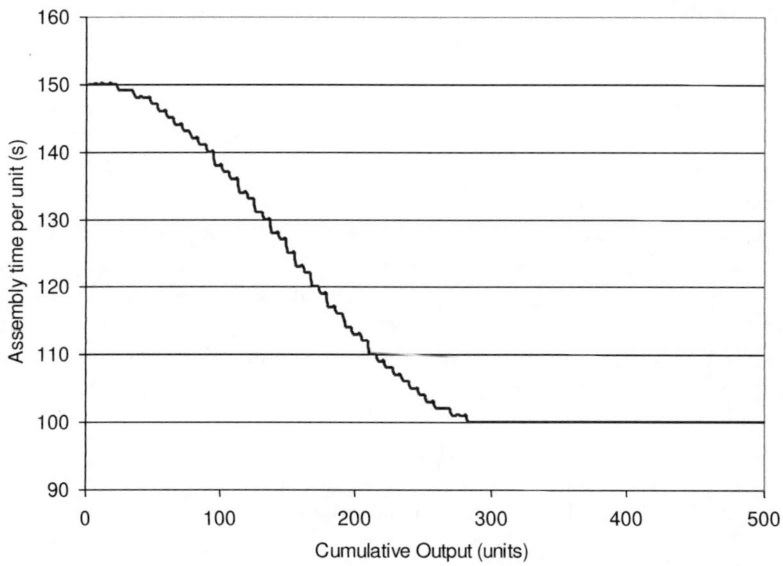

Figure 2. The behaviour of the learning curve during a learning process in assembly.

Table 1 shows a summary of related human factors that may affect human performance against a number of performance measures on human-centred manufacturing systems through a literature study.

Table 1. Human factors vs human performance measures on a human-centred manufacturing systems.

Human factors	Human performance measures	Literatures
Physical work capacity	Cycle time	Galen et al [11]; Govindaraju et al [12]; Boenzi et al [13]
	Job satisfaction	Govindaraju et al [12]; Narahari et al [14]
Age	Physical workload	Shepherd et al [15]; Schibye et al [16]; Bridger et al [17]; Stead et al [18]; Zwick et al [6]; Kenny et al [10]
	Reaction time	Woodson et al [19]; Der et al [20]; Salvia et

		al [21]; Svetina et al [22]; Doroudgar et al [23]
	Learning and forgetting	Zwick et al [6]; Stanley et al [24]; Boenzi et al [13]
	General cognitive abilities	Warr et al [25]; Boenzi et al [13]; Stanley et al [24]; Llmarinen et al [9]
	Job satisfaction	Rhodes et al [26]; Drabe et al [27]; Clark et al [28]; Kumar et al [29]
Gender	Physical workload	Bridger et al [17]
	Reaction time	Der et al [20]; Adam et al [7]
Experience	Cycle time	Reagans et al [30]
	Throughput	Reagans et al [30]; Hunter et al [31]
	Learning and forgetting	Shafer et al [32]; Hunter et al [31]; Reagans et al [30]
	Physical capacity	Giniger et al [3]
	Reaction time	Ando et al [33]; Visser et al [34]
	General cognitive abilities	Giniger et al [3]; Llmarinen et al [9]; Boenzi et al [13]
Learning and forgetting	Cycle time	Falck et al [35]
	Throughput	Shafer et al [32]; Hunter et al [31]
Job satisfaction	Throughput	Rodriguez et al [36]
Circadian rhythms	Reaction time	Berg et al [8]

3. Analysis of the impact of human factors on human performance

Figure 3 shows a study based on findings of the literature review by ranking and selecting the most influential human factor against one of performance measures of a human-centred manufacturing system using the Gephi network analysis method.

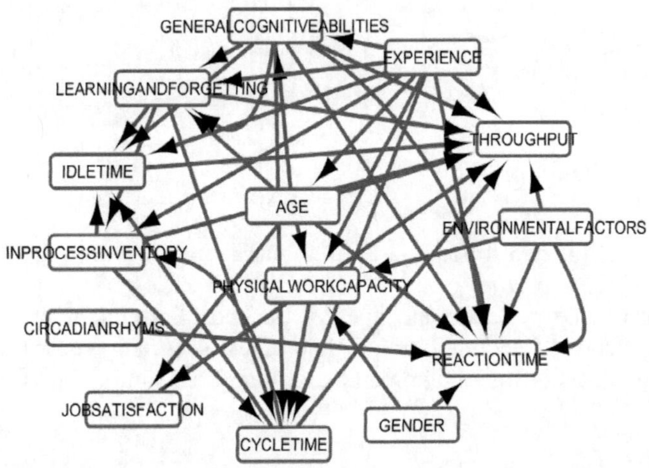

Figure 3. Importance of a human factor against one of performance measures using the Gephi method.

Figure 4 illustrates the result of impact rates of human factors in the form of eigencentrality on cycle time as one of performance measures using the Gephi network analysis method. It shows that experience is the most influential human factor that significantly affects cycle time of completing a task or product, compared to other human factors, although age and general cognitive ability also play a significant role on human performance. The result also indicates that circadian rhythm has the least impact on human performance. Interestingly, gender is the second least human factor that affects the human performance.

IOP Conf. Series: Materials Science and Engineering **239** (2017) 012006 doi:10.1088/1757-899X/239/1/012006

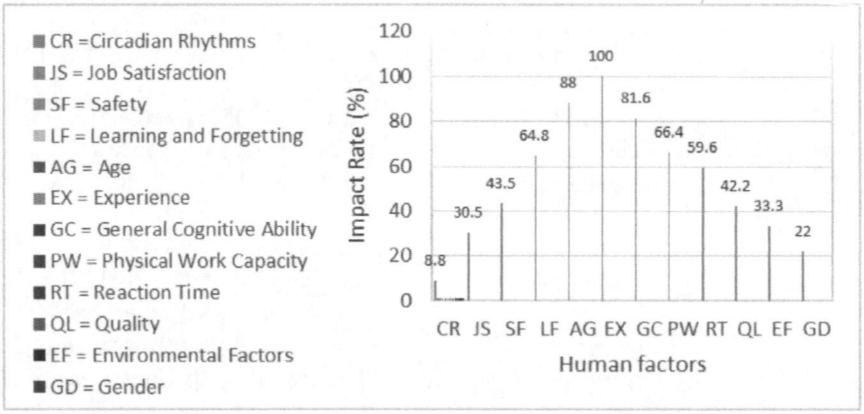

Figure 4. Impact rates of human factors on cycle time.

Figure 5 shows that age is the most influential human factor that affects throughput followed by experience and general cognitive ability. By comparison, circadian rhythm has the least impact on human performance and gender is the second least human factor that affects the human performance.

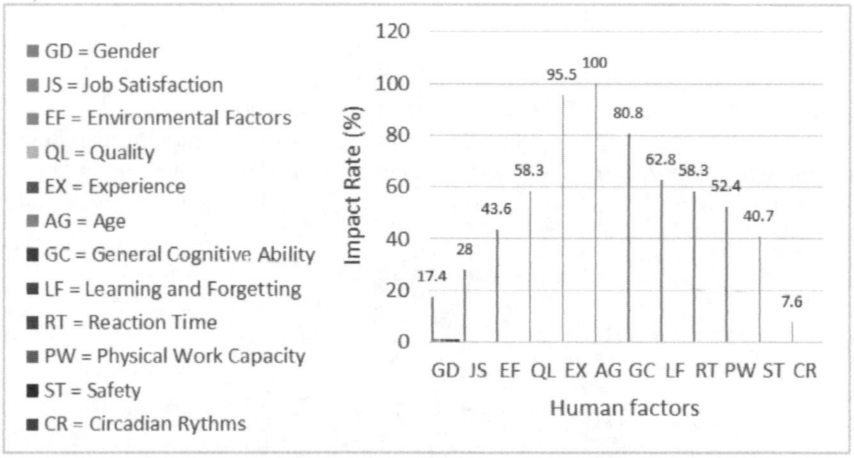

Figure 5. Impact rates of human factors on throughput.

Figure 6 shows that both experience and age are the most influential human factors on in-process inventory that also reflects the accumulation of idle times of human workers during production, while circadian rhythm is the least influential human factor that affects human performance.

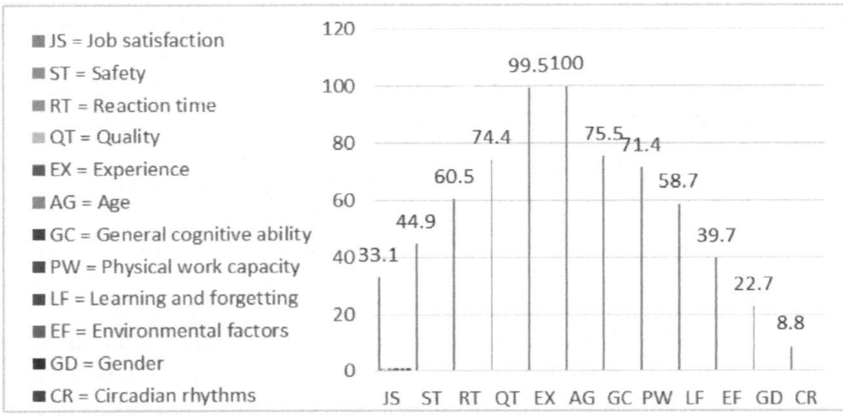

Figure 6. Impact rates of human factors on in-process inventory.

4. Conclusions and future work

The aim of this study was to identify human factors that affect human performance in human-centred manufacturing systems based on findings of a literature review. Much evidence from the literatures suggests that human performance is largely determined by experience rather than age in a view of socio-technical or psychological sciences. This study reports the initial research outcomes in terms of the interrelationship and the effects of human factors on human performance in manufacturing. The study also shows that both experience and age are the most significant human factors on human-centred manufacturing systems. In the latest developments of DES tools, although developers have been slowly introducing cloud-based technologies to facilitate the mobility of applications and the interoperability between different users or partners, none of these DES tools incorporate human factors as parameters used for manufacturing systems evaluation, which are still under developments. Research is required in order to develop more intelligent tools, autonomous and self-adapting systems for automated optimisation of system parameters.

References

[1] Wang Q, Sowden M and Mileham A R 2012 Modelling human performance within an automotive engine assembly line *Int J of Adv Manuf Tech 68* pp 141-148
[2] Wang Q and C R Chatwin Key issues and developments in modelling and simulation-based methodologies for manufacturing systems analysis, design and performance evaluation Int J of Adv Manuf Tech 25 pp 1254-1265
[3] Giniger S, Dispenzieri A and Eisenberg J 1983 Age, experience, and performance on speed and skill jobs in an applied setting *J of Appli Psych*, 68 pp. 469-475
[4] Avolio B J and Waldman D A 1990 Age and work performance in monmanagerial jobs: the effect of experience and occupational type *Acad of Manag J* 33 pp. 407 – 422
[5] McEvoy G M and Cascio W F 1989 Cumulative evidence of the relationship between employee age and job performance *J of Appli Psych* 74 pp. 11-17
[6] Zwick T and Gobel C 2009 Age and productivity-evidence from link employee data *Zew-centre for European Economic Research Discussion Paper*
[7] Adam J J, Pass F Bueke M J, Wuyts I J, Spijkers W and Wallmeyer P 1999 Gender differences in choice reaction time: evidence for differential strategies *Ergonomics* 42 pp. 327-335
[8] Berg J V and Neely G 2006 Performance on a simple reaction time task while sleep deprived *Percept & Motor Skills* pp. 589-599
[9] Llmarinen J E 2001 Aging workers *Occup and Env Med* 58 pp. 546-552
[10] Kenny P G, Groeller H, McGinn R and Flouris A D 2015 Age, human performance and physical employment standards *2nd Int Conf on Phy Employ Stand* pp 23-26
[11] Galen L 1987 Applied Ergonomics. Oxford: Butterworth

[12] Govindaraju M, Pennathur A and Mital A 2001 Quality improvement in manufacturing through human performance enhancement *Int Manuf Sys* pp. 360-367

[13] Boenzi F, Mossa G, Munmolo G and Romano V A 2015 Work Aging in production systems: modelling and performance evaluation. *25th DAAAM Int Sym on Intel Manuf & Auto* pp. 1108-1115

[14] Narahari C L and Koneru D K 2017. A study on the role of occupational stress in organisations. *Int J of Eng Tech, Eng, Manag & Appl Sci*, 5 pp. 53-59

[15] Shephard R J 2000 Aging and productivity: some physiological issues *Int J of Indus Erg* 25 pp. 535-545

[16] Schibye B and Hansen A F 2001 Aerobic power and muscle strength among young and elderly workers with and without physical demanding work tasks. *Appl Erg* 32 pp. 425-431

[17] Bridger R S 2009 Introduction to ergonomics United State of America: CRC Press

[18] Stead M and Page M 1983 Ergonomics at work England: Manpower Service Commission

[19] Woodson W, Tillman E and Tillman B 1992 Human factors design handbook: information and guidelines for the design of systems, facilities, equipment and products for human use In W. E

[20] Der G and Deary I J 2006 Age and sex differences in reaction time adulthood: result from the United Kingdom health and life style survey *Psych & Aging* 21 pp. 62-73

[21] Salvia E, Petit C, Champely S, Rienzo F D and Collect C 2016 Effects of age and task on drivers response accuracy and reaction time when responding to traffic lights, *Front in Aging Neuro*, pp. 1-9

[22] Svetina M 2016 The reaction times of drivers aged 20-80 during a divided attention driving *Traf Inj Prev* 17 pp. 810-814

[23] Doroudgar S, Chuang H M, Perry P J, Thomas K, Bohnert K and Canedo J 2017 Driving performance comparing older vs younger drivers. *Traff Inj Preven* 18 pp. 41-46

[24] Stanley R P 1985 Adult age differences in short term forgetting *Int J of Psycho 60 pp.* 83 - 101

[25] Warr P 1994 *Work and Aging: a European perspective.* Great Britain: Tailor and Francis Ltd

[26] Rhodes S 1983 Age-related differences in work attitudes and behaviour *Psych Bulleting*, 93 pp. 328-367

[27] Drabe D, Hauff S and Richter N F 2015 Job satisfaction in aging work force: an analysis of the USA, Japan and Germany. *Int J Human Res Manag* 26 pp.783-805.

[28] Clark A and Oswald A 1996 Is job satisfaction U-shape in age? *J of Occup & Organ Psych* 69 pp. 57-81

[29] Kumar T V 2017 Factors impacting job satisfaction among police personal in India: A multidimensional analysis Geogia: International criminal justice review

[30] Reagans R Argote L and Brooks D 2005. Individual experienced and inexperience working together: predicting learning rate from knowing who knows what and knowing how to work together. *Manage Sci* 51 *pp.* 869 - 881

[31] Hunter J E 1986 Cognitive ability, cognitive aptitudes, job knowledge, and job performance. *J of Voc Behav*, pp. 389-399

[32] Shafer M S, Nembhard A D and Uzumeri V M 2001 The effect of learning, forgetting, and heterogeneity on assembly line productivity *Manage Sci* 47 pp. 1639-1653

[33] Ando S, Kida N and Oda S 2004 Retention of practice effect on simple reaction time for peripheral and central visual fields *Erg* 98 pp. 897-900

[34] Visser I, Raijmakers M E and Molenaar P C 2007 Characterizing sequence knowledge using online measures and hidden markov models *Memo & Cog* 35 pp. 1502-1517

[35] Falck A C and Rosenqvist M 2012 What are the obstacles and needs of proactive ergonomics measures at early product development stages?-an interview in five Sweden companies Int J of Indus Erg, 42 pp. 406-415

[36] Rodriguez D, Buyens D and Landeghem H V 2016 Impact of lean on perceived job autonomy and job satisfaction: an experimental study *Human Factors & Erg in Manuf & Ser Indust*

IOP Conf. Series: Materials Science and Engineering **239** (2017) 012007 doi:10.1088/1757-899X/239/1/012007

Applying PCI in Combination Swivel Head Wrench

Tsang-Chiang Chen[1], Chun-Ming Yang[2], Chang-Hsien Hsu[3], Hsiang-Wen Hung[4]
[1]National Quemoy University, Kinmen County, Taiwan R.O.C
[2]Guilin University of Technology, Guilin, China
[3, 4]Asia University, Taichung, Taiwan R.O.C

Email: riarin04@gmail.com

Abstract. Taiwan's traditional industries are subject to competition in the era of globalization and environmental change, the industry is facing economic pressure and shock, and now sustainable business can only continue to improve production efficiency and quality of technology, in order to stabilize the market, to obtain high occupancy. The use of process capability indices to monitor the quality of the ratchet wrench to find the key function of the dual-use ratchet wrench, the actual measurement data , The use of process capability C_{pk} index analysis, and draw Process Capability Analysis Chart model. Finally, this study explores the current situation of this case and proposes a lack of improvement and improvement methods to improve the overall quality and thereby enhance the overall industry.

1. Introduction

Global competition in the era of intense competition and environmental changes, making Taiwan's traditional industries are facing economic pressures and industrial competition, the original Taiwan's traditional industries to a simple OEM production, gradually affected by globalization and lost the competitive advantage, today's traditional industries in Taiwan can Yong Continued operation to be diversified, automated, diversified services, and high-quality technology. However, under the current trend of demand for profitability, shortening of delivery and diversification of orders, only to continuously improve production efficiency and improve quality, and the establishment of flexible production capacity in order to compete in a fierce competitive environment, And achieved a high market share [1].

According to the report of the Taiwan Hand Tools industry [2], Taiwan's hand tools totalled NT $ 69.15 billion in 2014, about 2.2% higher than in 2013, and the total output value increased year by year. Billion, grew 18.61%(Figure1).

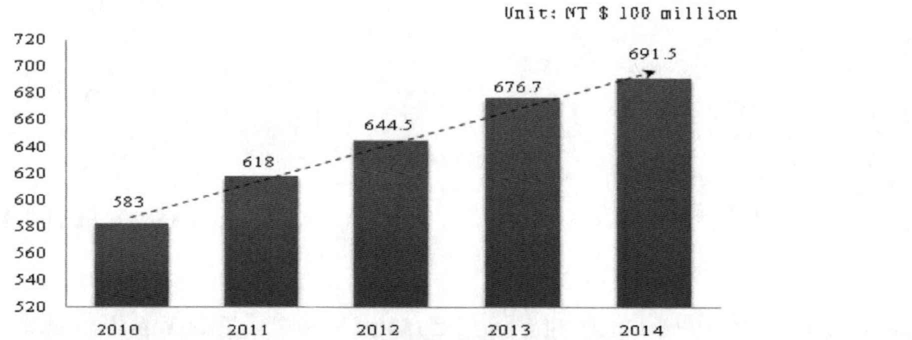

Figure 1. Taiwan hand tools total output value

In this study, Taichung City N Industrial Co., Ltd., the company mainly produces precision machining of metal parts, CNC lathe processing, meat grinder, food machinery parts, automobiles, locomotives and bicycle parts, rods and wrenches and many other products, Dual-use ratchet wrench for one of the products. In this study, we can improve the manufacturing process of N-use ratchet wrench with process capability index. We hope to improve the problem of poor quality, failure rate and quality management by means of process capability index. This study will discuss the current situation and propose a deficiency and improve the way to improve the overall quality, thereby enhancing the overall industry.

2. Literature

2.1. Quality Management.

The concept of quality management changes with the times, the use of the product point of view, including the characteristics of the product itself and the quality of the product; used in the process of the point of view, including the quality of technical processes [3]. In the process of use, [4] argue that quality management is the organization's commitment to continuous improvement through the culture of customer satisfaction, the two authors compare quality control between the Japanese company and the Japanese Original Will be affected by environmental characteristics, company type, company characteristics, and employee characteristics. Comprehensive quality management methods, are to effectively improve the quality and control.

2.2. Process Capability Index.

Process Capability Indices Originally proposed by [5], the concept of process capability indicators and ideas, when the product quality characteristics were stable and subject to normal distribution, the definition is:

$$C_p = \frac{USL - LSL}{6\sigma} = \frac{d}{3\sigma}$$

Where, USL for the process quality characteristics of the upper limit, LSL for the process quality characteristics of the lower specifications, the standard deviation of the quality characteristics, d for the deviation. Since C_p only considers the degree of process variation and ignores the mean value of the process, the [6] constructs C_{pk} with the process average as the center. In addition, he also suggests that in the unilateral rule, Looking for large process capability indicators C_{pu} and C_{pl}. Among them, C_{pk}, C_{pu} and C_{pl} of the expression is as follows:

$$C_{pk} = min\left\{\frac{USL-\mu}{3\sigma}, \frac{\mu-LSL}{3\sigma}\right\} = \frac{d-|\mu-m|}{3\sigma}$$

$$C_{pu} = \frac{USL-\mu}{3\sigma},$$

$$C_{pl} = \frac{\mu\ LSL}{3\sigma},$$

Which μ is the average of the process, m is the midpoint of the specification interval. Although Cpk can reflect the process mean (μ) deviation from the specification center (m), but can not reflect the deviation from the target T, and C_p when the degree of variation tends to 0, C_{pk} value will be divergent to infinity.

To measure a product usually has many processes, [7] defined the minimum individual PCI $C_0 = (1/3)\Phi^{-1}\left[(\sqrt[t]{2\Phi(3c)-1}+1)/2\right]$

, where c is the integrated PCI, t is the total number of process, Φ is the standard normal cumulative distribution function, and Φ−1 is the inverse of Φ.

3. Process capability index C_{pAP}

In this section, a new PCI C_{pAP} will be developed to measure the process capability of a product. Let the accuracy index A = μ/USL and the precision index P = σ/USL and using Equation (5), the new PCI CpAP can be presented as:

$$C_{pAP} = \frac{USL-\mu}{3\sigma} = \frac{1-A}{3P}.$$

Hence, the accuracy and precision of all the process of a product can be evaluated. Also, quality managers can use the new integrated PCI to identify the substandard quality characteristics of the product. To more effectively use the new PCI C_{pAP}, we establish four steps as follows.

Step 1: The total number of process t must be known and c must be decided. Then, C_0 can be determined.

Step 2: Collect data for each process.

Step 3: Calculate the C_{pAP} of all process using Equation.

Step 4: Identify the poor process in order to improve product quality and performance.

4. Numerical example

The tightening of the screws and nuts requires a wrench to adjust. If there is a difference in the quality of both ends of the ratchet wrench, it will make it impossible to fully interface with a variety of angles ratchet, error, yield can not be raised and other issues, easy to cause wrench fracture and easy to turn. So the quality of the wrench process is very important, the relationship between the dual-use ratchet wrench key factors.

In general, the process of the function or specification requirements must meet the product specifications, in order to ensure quality, such a process specification, we call the important "quality characteristics", therefore, this study to N company wrench process, for example, there are three important quality characteristics, namely C_{pkA}, C_{pkB}, C_{pkC} (Figure2).

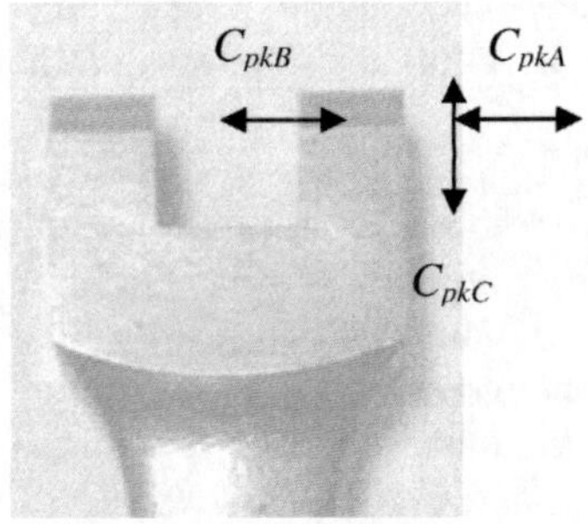

Figure 2. The reciprocating ratchet wrench has three quality features

As shown in Figure2, the two-purpose ratchet wrench has three "quality characteristics", namely: the right side of the wrench C_{pkA}, wrench left C_{pkB}, depth C_{pkC}, the three important quality characteristics of the process will be the focus, 3 important quality characteristics are looking for the purpose, so through the Institute of the development of quality assessment model, can identify the quality of the product quality of the important quality of the process, and improve the method and improve product quality and yield. The analysis section is as follows:

Step 1: the two-purpose ratchet wrench has three important quality characteristics, are looking for the head. Wrench on the right side of C_{pkA}, wrench left C_{pkB}, product specifications are 6.10 ± 0.15, depth C_{pkC} product specifications for 9.3 ± 0.2.

Step 2: 30 samples were taken by random sampling to calculate μ and σ, respectively.

Step 3: Calculate the three sets of process capability indicators by the formula Equation. The values of the USL are 6.25, 6.25 and 9.50 respectively, and the LSL values are 5.95, 5.95 and 9.10 respectively. The values of d are 0.15, 0.15 and 0.2, respectively, and the values of m are 6.10, 6.10,9.3.

$C_{pkA} = (0.15 - |6.10\text{-}6.10|) / (3 \times 0.03) = 1.67$

$C_{pkB} = (0.15 - |6.10\text{-}6.10|) / (3 \times 0.03) = 1.67$

$C_{pkC} = (0.2 - |9.56\text{-}9.30|) / (3 \text{ x } 0.09) = -0.22$

Step 4: Calculate the quality of the right C_{pkA} (1.67), the left side of the C_{pkB} (1.67) of the process capacity indicators have reached the quality level, are greater than 1; only depth C_{pkC} (-0.22) did not meet the process quality indicators of quality standards, it is clear wrench depth C_{pkC} of the accuracy and accuracy of the lack of quality improvement should be the process should be given priority to improve the process.

From the above steps, you can draw the PCAC mode diagram (Figure3):

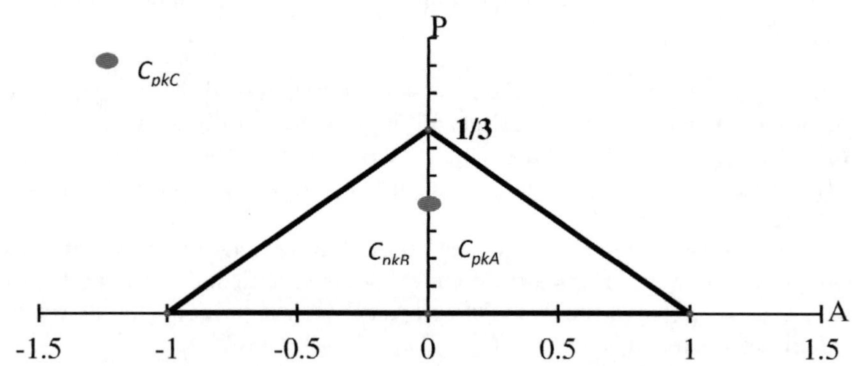

Figure 3. The dual-use ratchet wrench PCAC diagram

5. Conclusion

In this study, the case study method, N company's dual-use ratchet process, the use of process capability C_{pk} indicators to improve the process yield and improve product quality. Therefore, according to the present study, the main research results are as follows:

C_{pkA} (1.67) on the right side of the wrench, C_{pkB} (1.67) on the left side of the wrench, and the C_{pk} value is greater than 1. According to the corresponding relationship between the C_{pk} values of the different quality levels, the quality of C_{pkA} and C_{pkB} is excellent. Quality characteristic depth C C_{pkC} (-0.22) is -0.22, less than C_{pk} value of the standard value of 1, on behalf of the depth of the quality characteristics of C, need to improve the quality of the next step.

Discussion and discovery, the depth of C will be bad quality reasons, the processing conditions are not good, the power of the General Assembly variant and fracture, the depth is too deep, then easy to break with the ratchet and the lack of force, too shallow is not easy to turn.Further that the depth of C to improve the quality, need to replace the tool and knife mill, will make the company to increase costs and increase the burden on the company, so do not change. With the ratchet processing part of the downstream manufacturers, so this study is limited to the data cannot be measured ratchet, accurate measurement.

6. Acknowledgments

This work was supported by Research Fund of Guilin University of Technology of China.

7. References

[1] Pyng Sheu 2008 Feasibility of Implement Lean Production and Six Sigma in Small- and Medium-sized Enterprises-A Case Study of Taiwanese Small- and Medium-sized Electroplating Companies, Department Industrial Engineering and Management, Chaoyang University of Technology

[2] Metal Industry Research and Development Center 2015 104 years "global marketing, to create export dynamic marketing services train - hand tool industry": 2014 China's hand tool industry review and future prospects.

[3] Technical Department of the Ministry of Economic Affairs, Industrial Technology Knowledge Service Program 2015 2015 Yearbook of Metal Products - Hand Tools4.

[4] Kanji G K and Yui H 1997 Total quality culture. Total Quality Management,**8(6)**, 417-428.

[5] Juran J M 1974 Juran’s quality control handbook(3rded). New York:McGraw-Hill.

[6] Kane V E 1986 Process capability indices. Journal of Quality Technology, **18(1)**, 41-52.

[7] L Y Ouyang C H Hsu C M Yang 2013 A new process capability analysis chart approach on the chip resistor quality management, P. I. Mech. Eng. B- J. Eng. **227** 1075-1082.

IOP Conf. Series: Materials Science and Engineering **239** (2017) 012008 doi:10.1088/1757-899X/239/1/012008

Study of New Method Combined Ultra-High Frequency (UHF) Method and Ultrasonic Method on PD Detection for GIS

Yanran Li[1,a)], Duo Chen[1)], Jiwei Zhang[1)],Ning Chen[2)], Xiaoqi Li[1)], Xiaojing Gong[1)]

[1]Jinan Power-supply Company,Jinan, Shandong 250012, China;
[2]Zaozhuang Power-supply Company, Zaozhuang, Shandong 277100, China

[a)]lyr0021@yeah.net

Abstract. GIS (gas insulated switchgear), is an important equipment in power system. Partial discharge plays an important role in detecting the insulation performance of GIS. UHF method and ultrasonic method frequently used in partial discharge (PD) detection for GIS. It is necessary to investigate UHF method and ultrasonic method for partial discharge in GIS. However, very few studies have been conducted on the method combined this two methods. From the view point of safety, a new method based on UHF method and ultrasonic method of PD detection for GIS is proposed in order to greatly enhance the ability of anti-interference of signal detection and the accuracy of fault localization. This paper presents study aimed at clarifying the effect of the new method combined UHF method and ultrasonic method. Partial discharge tests were performed in laboratory simulated environment. Obtained results show the ability of anti-interference of signal detection and the accuracy of fault localization for this new method combined UHF method and ultrasonic method.

1. Introduction

GIS plays an important role in power system. Partial discharge is one of the important parameters to detect the insulation performance of GIS. It is recognized that partial discharge caused by free metal particles strongly harms the GIS system. The effects of UHF method and ultrasonic method on the partial discharge in GIS are currently of considerable interest. It is adopted to discriminate early insulation defects in GIS.

With the development of online PD detection, many papers have investigated the PD detection methods. S.Tenbohlen et al has reported a summary of PD measurements applying the UHF methods[1] . Simultaneously, Si et al has developed a data analyzer for ultrasonic detection system[2]. However, very few studies have been conducted on comparison and combination of this two methods. In this paper, the effects of the new method combined UHF method and ultrasonic method on PD detection in GIS were studied. The results revealed that can greatly enhance the ability of anti-interference of signal detection and the accuracy of fault localization.

2. Experiment

For UHF method, partial discharge in the GIS is accompanied by a very steep current pulse. Electromagnetic wave is propagated in the form of transverse electromagnetic waves (TEM) transmission, and high-order transverse electric field wave (TE) and transverse magnetic field (TM) are established[3].UHF antenna is used to detect electromagnetic waves generated by the partial discharge of GIS for the analysis of electromagnetic wave[4].For ultrasonic method. The molecules are violently collided with each other and pressure is created in the moment of partial discharge in GIS. Ultrasonic pulses includes longitudinal wave, shear wave and surface wave. The ultrasonic signal can be collected by the installed ultrasonic sensor in the GIS wall, conversed to electrical signals, and then the partial discharge level within the GIS can be measured.

UHF and ultrasonic methods are non-contact measurement methods. On-line PD detection can be achieved based on UHF and ultrasonic methods. Ultrasonic detection is suitable for field testing, while the UHF method can be used for continuous on-line monitoring. More complex equipment is required for ultrasonic method and higher cost is required for UHF method.

Considering the advantages and disadvantages of UHF method and ultrasonic method, a new method combined UHF method and ultrasonic method is proposed. This method has mixed distributed sensor technology. Obtained results show strong anti - interference ability, high positioning accuracy and high reliability of this method.

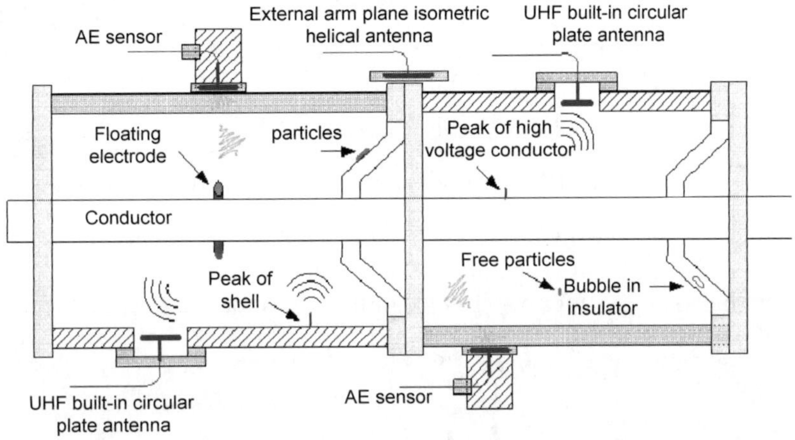

Figure 1. Schematic diagram of new combined method.

3. Analyzation

This new method for PD source detection is mainly based on the statistical characteristics and phase analysis of PD signal. It is a common pattern recognition problem. It can be carried out by the expert system combined with fuzzy logic, neural network (NN) and fractal analysis. The experimental specimens were made in the laboratory simulation environment. The typical UHF signal waveforms produced by free particles in GIS are measured.

In the laboratory simulation environment, the typical UHF signal waveforms produced by various typical defects are measured. Due to the limitation of space, the author only selects the UHF signal waveform caused by the discharge of the metal protrusion on the conductor. Waveform characteristics is extracted to establish a defect database and provide the basic data for the application of the expert system, as is shown in Fig.2.

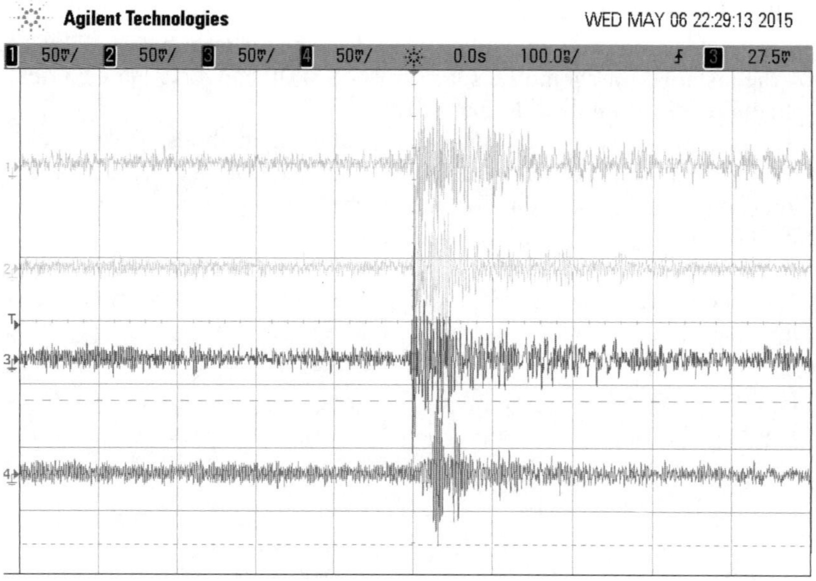

Figure 2. UHF signal wave form caused by free particles.

For this new method, a number of ultrasonic typical signals are produced by different insulation defects, It is showed the different types of defects in the ultrasonic typical signal in Fig.3[5]. The pattern recognition method is used for pattern recognition based on the artificial neural network.

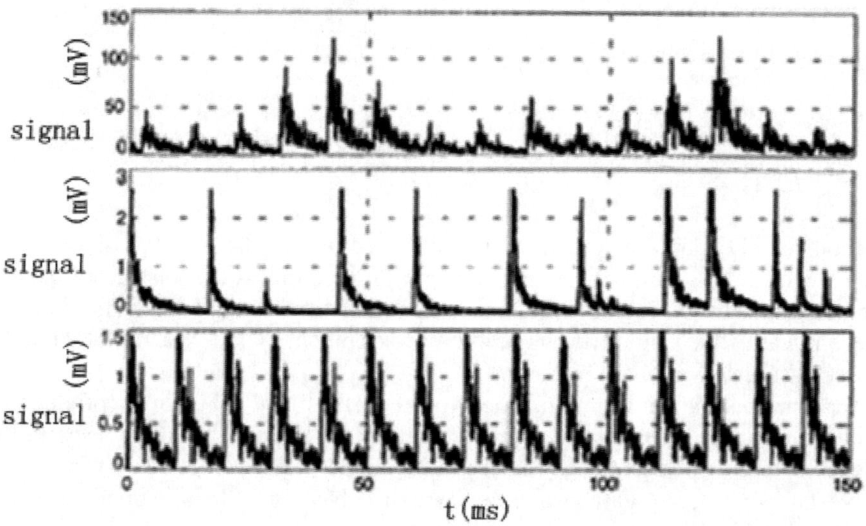

Figure 3. d) Moving particles e) Beating particles f) Loose screws

The partial discharge characteristic fingerprints can be constructed from the fingerprint parameters for the partial discharge phase distribution. Fingerprints and maps are an effective basis for judging the type and severity of discharges, and the use of maps can visually characterize the partial discharge profile.

A number of classical characteristic parameters are in the partial discharge analysis[6]. These statistical parameters can be used to construct partial discharge characteristic maps, as is shown in Fig.4.It is effective basis for judging the type of partial discharge and the severity of partial discharge.

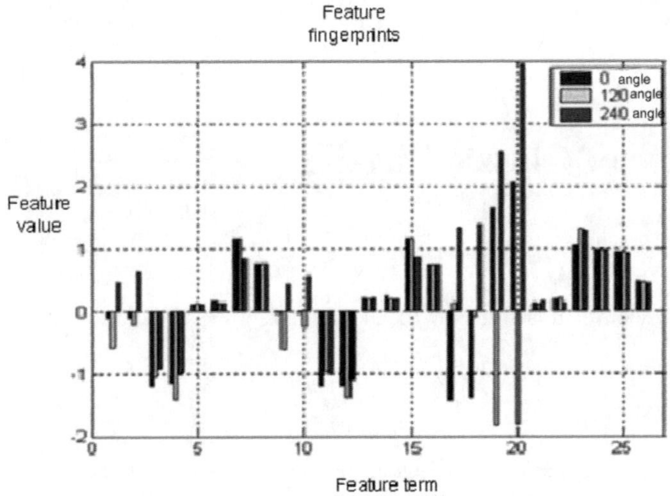

Figure 4. Partial discharge characteristic maps.

4. Conclusion

Considering the advantages and disadvantages of UHF method and ultrasonic method, a new method combined UHF method and ultrasonic method is applied in the partial discharge detection. Advantages of UHF method and ultrasonic method are mixed in this new method. It has mixed distributed sensor technology, strong anti - interference ability, high positioning accuracy and high reliability.

In our research, we found that on-line PD detection can be achieved by this new method combined UHF method and ultrasonic method. Its key technology is UHF and ultrasonic sensors. The ability of anti-interference of signal detection and the accuracy of fault localization can be greatly enhanced.

References

[1] S. Tenbohlen et al, Partial discharge measurement in the ultra high frequency (UHF) range[J]. IEEE Transactions on Dielectrics and Electrical Insulation, 2008. 15(6): 1544-1552.

[2] W. R. Si et al, Investigation of a comprehensive Identification Method used in acoustic detection system for GIS[J]. IEEE Transactions on Dielectrics and Electrical Insulation, 2009. 17(3): 721-732.

[3] U. Schichler, W. Koltunowicz, D. Gautschi, et al. UHF partial discharge detection system for GIS: Application guide for sensitivity verification: CIGRE WG D1.25[J]. IEEE Transactions on Dielectrics and Electrical Insulation, 2016, 23(3): 1313-1321.

[4] W. Gao, D. Zhao, D. Ding, et al. Investigation of Frequency Characteristics of Typical PD and the Propagation Properties in GIS[J]. IEEE Transactions on Dielectrics and Electrical Insulation, 2015, 22(3): 1654-1662.

[5] C. Boya, M. V. Rojas-Moreno, M. Ruiz-Llata, et al. Location of partial discharges sources by means of blind source separation of UHF signals[J]. IEEE Transactions on Dielectrics and Electrical Insulation, 2015, 22(4): 2302-2310.

[6] T. Li, X. Wang, C. Zheng, et al. Investigation on the Placement Effect of UHF Sensor and Propagation Characteristics of PD-induced Electromagnetic Wave in GIS Based on FDTD Method[J]. IEEE Transactions on Dielectrics and Electrical Insulation, 2014, 21(3): 1015-1025.

IOP Conf. Series: Materials Science and Engineering **239** (2017) 012009 doi:10.1088/1757-899X/239/1/012009

Quantum Metric of Classic Physics

Eugene Machusky

National Technical University of Ukraine "Kyiv Polytechnic Institute"

sivera@ukr.net

Abstract. By methods of differential geometry and number theory the following has been established: All fundamental physical constants are the medians of quasi-harmonic functions of relative space and relative time. Basic quantum units are, in fact, the gradients of normal distribution of standing waves between the points of pulsating spherical spiral, which are determined only by functional bonds of transcendental numbers PI and E. Analytically obtained values of rotational speed, translational velocity, vibrational speed, background temperature and molar mass give the possibility to evaluate all basic quantum units with practically unlimited accuracy. Metric of quantum physics really is two-dimensional image of motion of waves in three-dimensional space. Standard physical model is correct, but SI metric system is insufficiently exact at submillimeter distances.

1. Introduction

New SI measurement system will be introduced at the end of the year 2018 and will be based on constants derived from nature only [1]. So far, the main problem with regards to defining constants is finding the coordinated values of basic quantum units [2]. We show below that this problem can be solved not only without use of artifacts (meter, kilogram, second) but without any measurements at all. Comparison of analytically obtained units with their recommended by CODATA values is presented below:

Table 1. Comparison of measured and analytically obtained units

Unit	Measured (conventional)	Analytical
Speed of light	299792458	299792457.86759134
Background temperature	2.725	2.7252543275634558
Relative molar mass	0.012	0.0119992777505492
Kelvin	2.7316	2.7315999984590452
Fine structure	0.0072973525664(17)	0.0072973525205056
Avogadro	6.022140857(74)	6.0221410564201849*10^23
Boltzmann	1.38064857(74)	1.3806484502310000*10^-23
Planck	6.626070040(81)	6.6260700111158522*10^-34
Elementary charge	1.6021766208(98)	1.6021766150248797*10^-19
Newtonian gravitational	6.67408(31)	6.6740529112548490*10^-11

All analytically derived values lie within measurement accuracy considering the conventionally established values of speed of light and molar mass.

2. Analytic base of quantum physics

It is shown below that only five equations determine the analytic base of quantum physics.
Two equations constitute basis of electrodynamics of Maxwell-Gauss:

C = (R + 4*PI*C/10^18) ^64*10^7 = 299792457.867591338433684,
R = Integer[10^8*(C/10^7) ^ (1/64)]/10^8 = 1.05456978.

The number C is the value of rotational speed of radius-vector. The number R is the harmonic value of the inverse squared radius. The value 4*PI is solid 3D-angle in radians. The value 4*PI*10^-7 is the Magnetic constant. The value 4*PI*C*10^-7 is the Impedance of free space. The value 1/(4*PI*10^-7)/C^2 is the Electric constant. The number C is the Speed of light in physics, but arithmetically it is simply the number of turns around a ball with radius 4*PI/10^18 per unit of time. The value 1/10^18 is a natural limit of accuracy of calculations of an inverse squared radius due to finite numerical length of rotational speed.

Three equations constitute basis of thermodynamics of Kelvin-Avogadro:

K = E + AS + BS = 2.7315999984590452.
AS = 1/100/Sum {[137 + (137 - 100) *n]/10 ^ (3*n + 2)} = 0.00729 = 1/100/ (1.11111111111...) ^3.
BS = Sum [602214183/10^ (3*n + 8)] = 0.00602817 = 0.00602816999999...999999397183.

The number K is an upper limit of inverse temperature (used in SI for calibration of the triple point of water as 100*K = 273.16). The number E is the base of natural logarithm in mathematics but in physics it is a lower limit of inverse temperature. The word "temperature" means literally "rating of tempo" or, in other words, rating of a speed of vibrations. Instant value of translational velocity oscillates between the values of rotational and vibrational speeds [3]. The digital sequences AS = 0.00729 and BS = 0.00602817 have linked the alphabet (0...1) and the alphabet (0...9) because 1 = 0.999999..., and 10 = 9.99999... The digital sequence B = 602214183 is the Avogadro's integer. NB = B/(1+4*PI/10^8)/10^11 binds numbers B and PI. The sequence E = 2.7182818284590452 = (1+1/n) ^n exponentially connects 1 and infinity (by the hyperbola 1/n). The sequence PI = 3.1415962535897932 = Sum from (-1) to (+1) of [dx/sqrt(1-1/x^2)] links 1 and infinity (by the parabola x^2). Digital sequence A = 136.997250372498 connects PI and E because E^A = PI*10^59. Numbers PI and E are the absolute units of Space and Time.

3. Inverse geometry

The mutual bonds of geometrical parameters of pulsing sphere can be described by the following expression:

[G] = 2*PI*[R]*{1 + [A]},

where [R] is a matrix of relative inverse radii, [A] is a matrix of relative inverse eccentricities, 2*PI*[R] = [P] is a matrix of relative inverse perimeters. The instant values of radius Ri and eccentricity Ai are linked to each other by the parameter sqrt(2*PI*E) of Gaussian function of normal distribution [4; 5]:

Ri = 1 + 2/100*[E + Ai*(1 + sqrt(2*PI*E/100)],
Ai = [100*(Ri - 1)/2 - E]/ [1 + sqrt(2*PI*E/100)].

The values of inverse eccentricities are concentrated near reciprocal value of squared sum of mean values (root mean square, arithmetical mean, geometrical mean and harmonic mean) of numbers PI and E:

1/ {sqrt [(PI^2 + E^2)/2] + (PI + E)/2 + sqrt(PI*E) + 2*PI*E/ (PI + E)} ^2 = 1/136.9938985020083597.

The bond exponential of PI and E was illustrated by the expression PI*10^59 = E^136.9972503724980956.
Averages of inverse values of these digital sequences begin to differ from each other starting from the 12-th decimal place [6-8]. This is the natural limit of accuracy of inverse eccentricity calculations in polar coordinates.

3.1. Calibrating Fields

There are three fields of calibration of wave fronts. Fields of rotational and vibrational speeds are generating the field of translational velocity.

Matrix of rotational speed [CN] = (R + N*PI*[CN]/10^18) ^64*10^7 where N = 0, 1, 2, 4.

TABLE 2. Field of rotational speed

Notation	Value	Description
C4	2.9979245786759134*10^8	3D-speed (sphere)
C2	2.9979242359665663*10^8	2D-speed (circle)
C1	2.9979240646119366*10^8	1D-speed (diameter)
C0	2.9979238932573362*10^8	0D-speed (point)

TABLE 3. Field of vibrational speed

Notation	Value	Description
TE	2.9979245625727419*10^8	(R+1/E/10^8) ^64*10^7
	2.9979245616776434*10^8	Median (TA; TE)
T	2.9979245608618159*10^8	Harmonic temperature (median of two medians)
TA	2.9979245607825451*10^8	[R+1/(E+AS)/10^8] ^64*10^7
	2.9979245600459885*10^8	Median (TA; TK)
TK	2.9979245593094320*10^8	[R+1/(E+AS+BS)/10^8] ^64*10^7

TABLE 4. Field of translational velocity

Notation	Value	Description
C	2.9979245786759134*10^8	Harmonic rotational speed (3D-speed)
	2.9979245697688650*10^8	Root mean square
	2.9979245697688646*10^8	Arithmetical mean
V	2.9979245697688647*10^8	Harmonic translational velocity (median)
	2.9979245697688646*10^8	Geometrical mean
	2.9979245697688647*10^8	Harmonic mean
T	2.9979245608618159*10^8	Harmonic vibrational speed (temperature)

An arithmetical mean above is less than harmonic mean. This is impossible in classic analysis. This fact shows information entropy of discrete analysis and gives the key to calibration of mathematical 2D-image of 3D-motion of wave front.

3.2. Fields of Geometrical Parameters

The inverse eccentricities are concentrated near the number A1 = 1/137 = 0.007299270072992700729927... The unique endless decimal fraction A1 = Sum [729927/10^(8*n)] has a mirror symmetry with period 10^8 and in pair with AS = 1/Sum {[137 + (137 - 100) *n]/10^(3^n)} = 0.00729 determines the limits [9] of the field of inverse geometrical parameters compressed around the central value A0 = (PI*E/100) ^2 = 0.0072927060593902. The corresponding radii are as follows:

TABLE 5. Field of external radii

Notation	Value	Description
R1	1.0545719538152265	1 + 2/100*(E + A1*(1 + sqrt(2*PI*E/100)))
R01	1.0545718610477836	Median (R0; R1)
RP	1.0545717955420578	Median of two medians (harmonic Planck's radius)
R0S	1.0545717300363340	Median (R0; RS)
RS	1.0545716917923240	1 + 2/100*(E + AS*(1 + sqrt(2*PI*E/100)))

P01 = 6.62607042270046640.
2*PI*(R0S; R01) = 6.62607001111585250 = PP.
P0S = 6.62606959953125120.

Values of radius RP and perimeter PP have been identified as the harmonic Planck's units.

TABLE 6. Gauge field of radii

Notation	Value	Description
RC	1.0545697837673031	(C/10^7) ^ (1/64)
RV	1.0545697837183468	Median (RT; RC)
RE	1.0545697836787944	R + 1/E/10^8
RAE	1.0545697836738746	Median (RA; RE)
RT	1.0545697836693905	Median (RAK; RAE)
RA	1.0545697836689549	R + 1/ (E + AS)/10^8
RAK	1.0545697836649065	Median (RA; RK)
RK	1.0545697836608581	R + 1/ (E + AS + BS)/10^8

Background temperature:

Median (E+AS; K) = 2.7285850810946404. It is upper limit of TBG.
Median of medians = 2.7252543275634558 = TBG. It is harmonic background temperature.
Median (E+AS; E) = 2.7219256081809304. It is lower limit of TBG.

Limits of gauge eccentricity:

Lower limit A(RT) = 0.0072224927756108 = AT.

Upper limit A(RC) = 0.0072224962396475 = AC.

Molar amplitude of translational velocity:

Upper limit = 0.01199927775072243892.
[12 - (AT; AC)/10]/1000 = 0.01199927775054923709 = MM. It is harmonic molar mass unit.
Lower limit = 0.01199927775037603525.

Molar phase of translational velocity:

1.3806484502840000.
cos [12 - (AT; AC)/10] - sin [12 - (AT; AC)/10] = 1.3806484502310000 = KB. It is harmonic Boltzmann's unit.
1.3806484501770000.

TABLE 7. Internal radial field

Notation	Value	Description
RS	1.0545716917923240	R(AS); AS = 0.00729
	1.0545707377551196	Median (RV; RS)
RV	1.0545697837183468	(V/10^7) ^ (1/64)
	1.0545670092049269	Median (RV; RX)
RX	1.0545642346951568	R(AX); AX = 5/X – 1

Median of two above medians 1.0545688734791990 = RM is the harmonic value of internal radius.
X = Root [X*E^X/ (E^X - 1) = 5] is parameter of expression for Wien wavelength displacement.

Fine structure constant is determined as the eccentricity AF connecting numbers 137 and PI:
AF = 1000/ Integer [sqrt (137^2 + PI^2) *1000] = 0.0072973525205056.
Orbital radius:

RF = 1.0545718996147182.
Median (RF, RP) = 1.0545718475783874 = RQ. It is harmonic value of orbital radius.
RP = 1.0545717955420578.

Orbital density of velocity field:

1.6021766174049711.
Sqrt [AF*RP/ (T; C)] = 1.6021766150248797 = Q. It is the Elementary charge unit.
1.602176612644788.

Harmonic median density of velocity field:

6.6744109106473266.
2*PI*(RM; RQ) *[1 + (AM; AQ)] = 6.6740529112548490 = G. It is the Newtonian gravitational unit.
6.6736948926580828.

Entropy of calculations of geometrical parameters has been determined using the expressions connecting product PI*E and ratio 1/137:

IOP Conf. Series: Materials Science and Engineering **239** (2017) 012009 doi:10.1088/1757-899X/239/1/012009

[Ni] = 100*{sqrt [8*PI*E/ (137^2 + 8*PI*E)]/ (1 + 2*[Ai]/1000) - 5/10^8}.
P1 – P0 = 4*PI*10^-7*[2*E/ (PI + E) – cos(PI/6) *10^-4 – tan(PI/4) *10^-7 + sin(PI/6) *10^-9].

TABLE 8. Entropy of eccentricities

Eccentricity	Value	Description
AA	0.0073189621138002	(PI*E/100) ^2 + 4*[1/137 - (PI*E/100) ^2]
	0.0073188455271700	Median (AD; AA)
AD	0.0073187289405399	1/(16*PI*E)
AN	0.0073158881634932	Median of two medians
	0.0073129313978999	Median (AD; AL)
AL	0.0073071361524362	1/[ln(E) + 59*ln (10)]

Harmonic entropy unit:

NB = 6.0221410732354338 = B/ (1 + 4*PI/10^8)/10^8.
Median [NB; N(AN)] = 6.0221410564201849 = NA. It is harmonic Avogadro's unit.
N(AN) = 6 0221410396049361.

3.3. Decimal Orders of Quantum Constants
The decimal orders of the quantum constants are determined from the equations of standard physical model using exponential bond expression E^137 = 100*PI*10^57. For NA it is 10^ +23. For KB it is 10^ -23. For RP and PP it is 10 ^ - 34. For Q it is 10^ - 19. For G it is 10^ + 11. More information could be found in references [10 - 12].

3.4. Harmonic Symmetry of Quantum Physics
All quantum units are the parameters of density of the field of progressive velocity. Velocity is the relative quantity. The number C is the upper limit of progressive velocity. The harmonic velocity V was determined as the ratio of units of relative space and relative time. All quantum constants are relative units due to relativity of space and time. Harmonic symmetry of quantum physics is illustrated by Table 9.

TABLE 9. Quanta of energy

Energy quanta	Description
m*v^2/2*[1 + v^2/ (T…C) ^2]	From (m*v^2/2) to (m*C^2) when v approaches C; (Newton, Einstein)
h*f*[1 + v^2/ (T…C) ^2]	From (h*f) to (h*f*2) when v approaches C; (Kotelnikov, Nyquist)
k*T/ [1 + v^2/ (T…C) ^2]	From (k*T) to (k*T/2) when v approaches C; (Boltzmann, Shannon)
q*U/ [1 + v^2/ (T…C) ^2]	From (q*U) to (U*q/2) when v approaches C; (Dirac, Kaufmann)

where m, v, T, h, f, k, q, U are the instant values of mass, velocity, temperature, action, frequency, phase, electric charge and voltage.

4. Conclusions
Quantum Physics is Differential metric of pulsing spherical spiral.
Quantum Physics is digital bridge between Discrete and Continual Mathematics.
Quantum Physics is Absolute calculus of Wave motion.

5. References
[1] Proposed redefinition of SI base units/ en.wikipedia.org/.
[2] On the future revision of the SI/ www.bipm.org/.

[3] Equation of motion/en.wikipedia.org/.

[4] Machusky E.A. Logarithmetics of quantum physics/ 16-th Int. Scientific M.Kravchuk Conf., May 14-15, 2015, Kyiv: Conf.materials. Vol. 2. Algebra. Geometry. Analysis. – K.: NTUU "KPI", 2015. – pp.26 – 29.

[5] Мачуський Євген. *Альфа-PI-E гармонія природи*/ ISSN 1819-7329. "Світогляд", 2015, №2 (52), p.3.

[6] Zhivkov A.P. *On some numerical properties of the fine structure constant*/ A.P.Zhivkov, E.A.Machusky, A.V.Frolov/ Математика в сучасному технічному університеті: матеріали III Міжнар. наук.-практ. конф. (Київ, 25 – 26 груд. 2014 р.). - Київ: НТУУ «КПІ», 2015. – сс. 5 – 7.

[7] Мачуський Є.А. Mystery of 137 prime or simplex quantum theory/ Інформаційна безпека УкраЇни: Зб. наук. доп. та тез наук.-техн. конф; м.Київ, 12 - 13 березня 2015 р., Київський національний університет імені Тараса Шевченка/ Редкол.; Анісімов (голова) та ін.. – К.: Київський національний університет імені Тараса Шевченка, 2015. – сс. 16 – 18.

[8] Machusky E.A. Stairsing of 137 prime as a star sing of nature/ the same conference/, pp. 62 – 63.

[9] Machusky E.A. Means' subordination breaking by the means of means of Alpha-trinity/ E.A. Machusky, S.S. Mikadze/ Математика в сучасному технічному університеті: матеріали IV Міжнар. наук.-практ. конф. (Київ, 24 – 25 груд. 2015 р.). – Київ: НТУУ «КПІ», 2016. – P.10 – 12.

[10] Machusky E. Information security as base of General Safety and Security/ Information Security – International Training Workshop: Kyiv, Editor V. Korsun. – Kyiv: NTUU "KPI", 2014 – pp. 25 – 33.

[11] Мачуський Є.А. Диференціальна альфа-система числення в обернених кільцях/ Тези доп. V Міжнар. наук. -практ. конф. «Методи та засоби кодування, захисту й ущільнення інформації» м.Вінниця, 19 – 21 квітня 2016 р. – Вінниця: ТОВ «Нілан – ЛТД», 2016. – сс. 143 – 145.

[12] Мачуський Є. «Електрон»: науково-популярне видання/ ISBN 978-966-622-572-9. - Мачуський Є., - К.: НТУУ «КПІ», 2013. -32с.

IOP Conf. Series: Materials Science and Engineering **239** (2017) 012010 doi:10.1088/1757-899X/239/1/012010

Study a Fluid Structure Interaction Mechanism to Find Its Impact on Flow Regime and the Effectiveness of This Novel Method on Declining Pressure Loss in Ducts

Hamidreza Kamali[1*], Masoud Javan Ahram, S. Ali Mohammadi

Faculty of Mechanical Engineering, Islamic Azad University, Bushehr, Iran

[1]Mechanicengineer1369@gmail.com

Abstract. Using channels and tubes with a variety of shapes for fluids transportation is an epidemic approach which has been grown rampantly through recent years. In some cases obstacles which placed in the fluid flow act as a barrier and cause increase in pressure loss and accordingly enhance the need to more power in the entry as well as change flow patterns and produce vortexes that are not optimal. In this paper a method to suppress produced vortexes in two dimension channel that a fixed square cylinder placed in the middle of it in Re_D 200 in order to find a way to suppress vortexes are investigated. At first different length of splitter plates attached to square obstruction are studied to obtain the effects of length on flow pattern. Subsequently simulations have been conducted in three dimension to validate previous results as well as acquire better understanding about the selected approach. Simulations have done by Lagrangian Eulerian method, plates first assummed fix with length 1.5mm, 4mm and 7.5mm, and then flexible plates with the same length are studied. Young's modulus for flexible plate and blockage ratio were constant values of 2×10^6 and 0.25 in all simulations, respectively. Results indicate more vortexes would be suppressed when the length of splitter plate enhances.

1. Introduction

Transporting fluids using with high efficiency is an issue that has been inevitable nowadays due to conditions such as raising energy and equipment costs. Many companies in gas and oil industry try new kind of accessories, for example network of pipes with the least waste, to transport the raw material they required as well as their products with minimum cost and the least pressure loss in fluid flow beside maximum safety as the major goals of substance transportation. However using curve shapes in these systems, as said before, and subsequently vortex production because of facing fluids with walls or obstructions is unavoidable. Additionally, in many cases such as suspended bridges and also underwater structures vortex generation can be more harmful and cause numerous destructions, ruining Tacoma Narrow bridge because the blowing of a light wind is a good example to show the destructive effetc of this phenomenon. Thus finding a way to suppress vortices has became neccessary and various research groups from all over the world are trying to counteract this fact using a wide range of novel ideas.

In this paper our endeavours specifically allocated to study the vortex generation patterns and the impact of attending splitter plate on flow. Fixed plates' capabilites in destroying vortices are compared flexible plates with the same length and results presented in figures that are drawn by Ansys 16.2 simulator. Obstacle blockage ratio has obvious impression on flow patterns as well as vortex generation. So we picked 0.25 and selected square cylinder as these selections preferred to other

IOP Conf. Series: Materials Science and Engineering **239** (2017) 012010 doi:10.1088/1757-899X/239/1/012010

shapes and values by researchers because of their efficiency. The cylinder positioned with a 45 degree rotation into its center to cause less pressure loss.[1-5]
Besides, This point that flexible structure vibrations intensify while a fluid with the same or approximately same natural frequency flow over it has been proved by researchers[6-8], both experimentally and numerically. So there is two ways in order to decrease the volume of destructions; change in flow pattern and charactristicts which is not convenience in many situations, or detecting the best length of splitter plate to diminish the number of vortexes.

2. Model description

In this investigation a channel just like figure 1. With presented geometry and a flow of air with mentioned charactristics are studied. A developed profile for velocity, with the maximum value of 2.38 m/s in th middle of profile, has defined in order to make it easy to investigare the laminar flow. Properties of air presumed to be constant through simulations and viscous dissipation is omitted. Therefor, governing equations for laminar flow (1-2), velocity profile (3), young's modulus of flexible structure (4), and pressure loss (5) can be written as followings;

$$\nabla . V = 0 \tag{1}$$

$$\rho(\partial v/\partial t) + \rho V . \nabla V = - \nabla P + \mu \nabla^2 V \tag{2}$$

$$\frac{u}{u_{\max}} = (1 - (1 - \frac{2y}{w})^2) \tag{3}$$

$$E = 0.9465(\frac{mf^2}{b})(\frac{L^3}{t^3})(1 + 6.585(\frac{t}{L})^2) \tag{4}$$

$$f = (\Delta P) / (\rho u^2) (L/D_h) \tag{5}$$

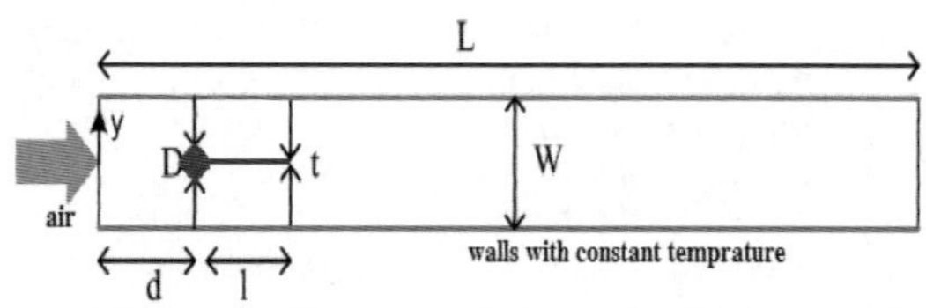

Figure 1. Shematic of channel with its attachments

Exact statictics about the geometry can be found in Table1. To mesh the fluid domain by picking Arbitrary Lagrangian Eulerian method as a renowned manner for its accuracy, the domain divided into 47814 cells. Furtheremore, the number of mesh in solid domain determined to be 1783 cell according to the feedbacks that we received from simulations. These simulations have conducted by coupling the Ansys fluent 16.2 and Ansys transient structural 16.2 while the time step for transfering data between the mentioned softwares was tuned to 0.01 second.

Table 1. Dimensions (mm) of the channel and structure

Lenghth of channel (L)	50mm	Diameter of square cylinder	1.5mm
Width of channel (w)	6mm	Distance between cylinder center and channel entrance	6mm
Thickness of Plate (t)	1mm		
Length of plate (l)	1/5, 4 and, 7.5mm	Angle between the square side and the channel surface	45^0

Nomenclature

ρ (Kg/m^3)..... Density	D (mm)............................ Diameter of cylinder
u(m/s)............ Fluid velocity	D_h (mm)...........................Hydrolic diameter
E(Pa)............. Young's modulus	m (kg).............................. weight
f(N)................ Friction	b (mm)............................ Depth
Re.................. Reynolds Number	Pr......................... .... Prandtl Number
ΔP.................. Pressure drop	L (mm)............................ Length of channel
W (mm)........Width of chan	t (mm)............................. Thikness of plate
d (mm)..........Distance from cylinder axis	l (mm)................................ length of plate

3. Results and Discussion

When the flow pass over the cylinder, or any kind of other obstructions, different kind of vortexes depend on Re number and other fluid features generate. To make this issue more clear and compare the attendance of fixed plate with different lengths, vorticity contours for fixed plates as well as emtpy channel and channel with fixed cylinder presented. Figure 3 illustrate this points clearly that attending splitter plate decrease number of vortices in fluid domain and this phenomenon intensify by increasing the length of plate.

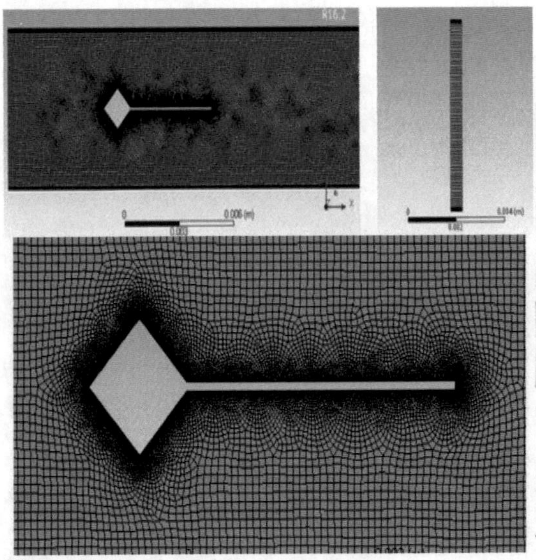

Figure 2. Fluid mesh

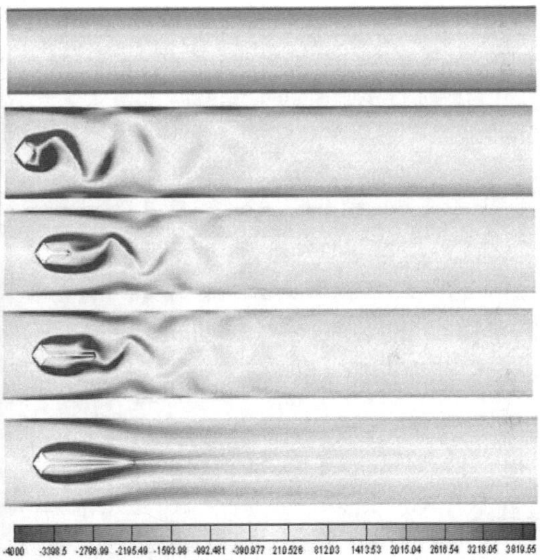

Figure 3. vorticity shematic for empty chnnel, Channel with cylinder, 1.5mm fixed plate, 4mm fixed plate, and 7.5mm fixed plate attached to cylinder

Pressure disturbution for 1.5mm, 4mm, and 7.5mm is drawn in figure 4. It shows that flexible plate has similar function to fixed plate in change flow regim as a splitter. While the length of plate growed we saw that vortexes declined. Another phonomenon that detected along with increase in the length of plate was vorex induced by vibrations of plate that were initialized by interactions between flexible plate and generated vortexes. This fact was amplitude by reactions of flexible plate to vortices. However, by enhancing the length of splitter the mentioned effect dwindled and the effect of length was dominant. The peak of vortex induced vibrations impacts diagnosed in flexible plate with 4mm length.

In the same way, drag and lift forces are compared for flexible plates in Table 3 and Table 4 respectively. As it can be viewed brightly, drag force has reverse correlation with the rise in the length of plate. Although the Lift force situation is a bit different and its quantity enhance simultaneously by length to 4mm, after this length the Lift declines constantly by increase in length of splitter plate. This is another evidence that proved the effect of attendance of flexible plate in quieting the flow regime.

IOP Conf. Series: Materials Science and Engineering **239** (2017) 012010 doi:10.1088/1757-899X/239/1/012010

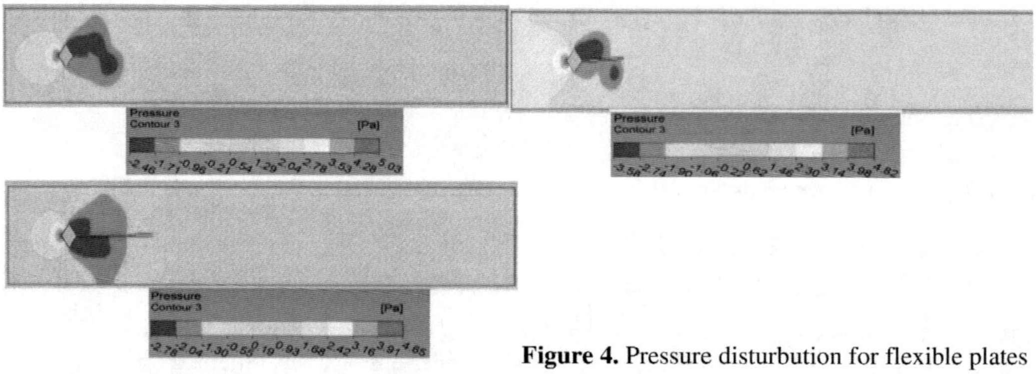

Figure 4. Pressure disturbution for flexible plates

Table 2. Drag for flexible plates

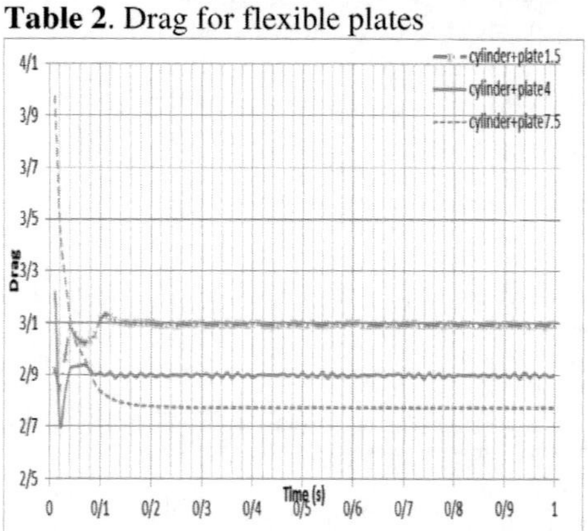

Table 3. Lift for flexible plates

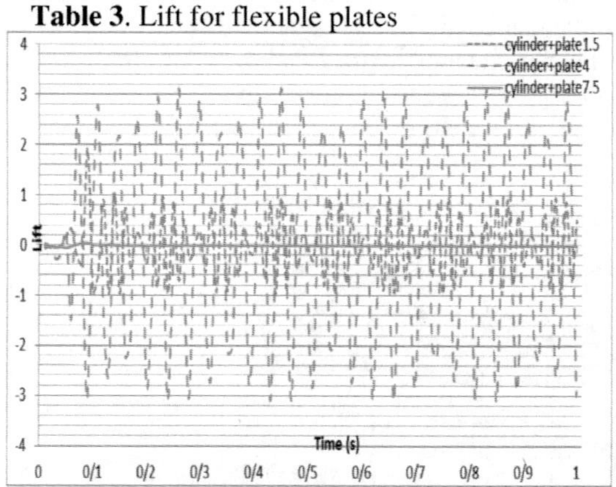

Table 4. Displacements of flexible plate in every second

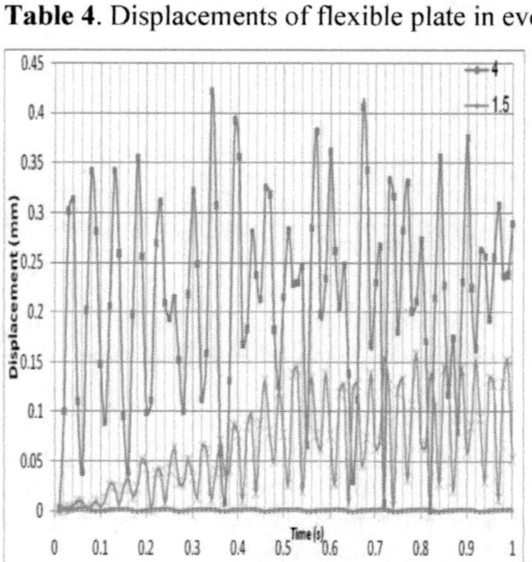

Table 5. Pressure loss comparison

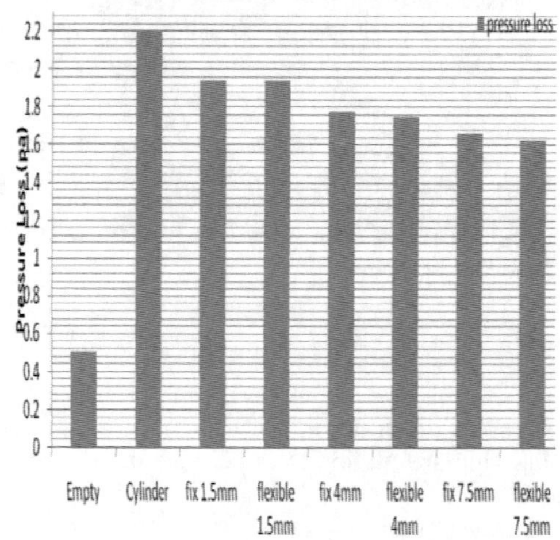

Table 4 illustrate the displacements of flexible plates in millimeter. Flexible plate with 4mm length had the most movements. The smallest plate was approximately fixed and had very little displacement.

IOP Conf. Series: Materials Science and Engineering **239** (2017) 012010 doi:10.1088/1757-899X/239/1/012010

The longest plate under investigation had movements less than 4mm plate but they were restricted and without any jump.

4. Conclusion

consequently pressure loss in all cases have been compared in Table 5. The biggest dissipation occurred in channel with cylinder situation. Attaching a plate, regardless to be flexible or fix, both make flow quiet and this effect strengthened by increase in the length of the plate. Flexible longer plates have better performance in relation to shorter ones. This phenomenon could be seen obviously in 7.5mm plate that its movement cause suppressing the vortexes.

5. Reference

[1] Aboueian-Jahromi, J., Nezhad, A. H., & Behzadmehr, A. (2011). Effects of inclination angle on the steady flow and heat transfer of power-law fluids around a heated inclined square cylinder in a plane channel. Journal of Non-Newtonian Fluid Mechanics, 166(23), 1406-1414.

[2] Kumar, A., Dhiman, A. K., & Bharti, R. P. (2014). Power-law flow and heat transfer over an inclined square bluff body: effect of blockage ratio. Heat Transfer—Asian Research, 43(2), 167-196.

[3] Srikanth, S., Dhiman, A. K., & Bijjam, S. (2010). Confined flow and heat transfer across a triangular cylinder in a channel. *International Journal of Thermal Sciences*, *49*(11), 2191-2200.

[4] Agarwal, R., & Dhiman, A. (2015). Confined flow and heat transfer phenomena of non-Newtonian shear-thinning fluids across a pair of tandem triangular bluff bodies. *Numerical Heat Transfer, Part A: Applications*, *68*(2), 174-204.

[5] Gomes, J. P., & Lienhart, H. (2013). Fluid–structure interaction-induced oscillation of flexible structures in laminar and turbulent flows. Journal of Fluid Mechanics, 715, 537-572.

[6] Griffith, M. D., Leontini, J., Thompson, M. C., & Hourigan, K. (2011). Vortex shedding and three-dimensional behaviour of flow past a cylinder confined in a channel. Journal of Fluids and Structures, 27(5), 855-860.

[7] Fujarra, A. L. C., Pesce, C. P., Flemming, F., & Williamson, C. H. K. (2001). Vortex-induced vibration of a flexible cantilever. Journal of Fluids and Structures, 15(3-4), 651-658.

[8] Srinil, N. (2010). Multi-mode interactions in vortex-induced vibrations of flexible curved/straight structures with geometric nonlinearities. Journal of Fluids and Structures, 26(7), 1098-1122.

IOP Conf. Series: Materials Science and Engineering **239** (2017) 012011 doi:10.1088/1757-899X/239/1/012011

The Vibration Analysis of Tube Bundles Induced by Fluid Elastic Excitation in Shell Side of Heat Exchanger

Minle Bao [1], Lu Wang[1], Wenyao Li[1] and Tianze Gao[1]

1 Key Laboratory of Ocean Energy Utilization and Energy Conservation of Ministry of Education, Dalian University of Technology, 116023, China
E-mail: baominle@hotmail.com

Abstract. Fluid elastic excitation in shell side of heat exchanger was deduced theoretically in this paper. Model foundation was completed by using Pro / Engineer software. The finite element model was constructed and imported into the FLUENT module. The flow field simulation adopted the dynamic mesh model, RNG k-ε model and no-slip boundary conditions. Analysing different positions vibration of tube bundles by selecting three regions in shell side of heat exchanger. The results show that heat exchanger tube bundles at the inlet of the shell side are more likely to be failure due to fluid induced vibration.

1. Introduction

Heat exchanger is one of the main industrial equipment which can improve energy efficiency and achieve heat exchange. It is widely used in various industrial fields, playing an important role in the production [1]. With the rapid development of industrial production and the continuous expansion of production scale, heat exchanger has been toward the development of large-scale and high parameterized [2-4]. But fluid induced vibration has become more and more obvious, because the non-supporting span of the heat exchanger bundles increases which has decreased the rigidity of the tubes. The failure of the heat exchanger bundles poses serious harm to the industrial production [5].

Fluid induced vibration incorporates a wide range of disciplinary knowledge, such as structural mechanics, material mechanics, fluid mechanics, and vibration. It is essential for the design and operation of heat exchangers. Up to now, scholars have done a lot of work around this discipline. At present, it is generally believed that the fluid-induced tube bundle vibration mechanism includes vortex shedding, turbulence buffeting, fluid elastic vibration, acoustic resonance and so on [6-7]. In this paper, we studied fluid elastic vibration in the shell flow field. In the case of the existence of fluid-induced vibration mechanisms, the vibration of any tube bundle changes the flow field around the tube bundle [8]. The change of the flow field causes the change of excitation force acting on the adjacent tubes. Fluid elastic vibration is caused by the interaction between the fluid exciting force and the elastic displacement of the tubes. The amplitude of heat exchanger tube bundles increases when the velocity of the fluid exceeds critical velocity of vibration induced by fluid elastic excitation. If the damping isn't large, the amplitude will increase until the tubes collide with each other [9-11].

Studies have shown that the fluid elastic vibration mechanism took place in the high velocity region mainly. It determines whether the fluid elastic vibration in the heat exchanger causes the tube bundle to vibrate in this paper. The dynamic mesh model in FLUENT software is used to determine the transverse displacement of each tube bundle.

2. Numerical approach

2.1. Mathematical models [12]

In this paper, Lever and Weave's model [13-14] of flow tubes are adopted to explain the instability of fluid elasticity. The assumption in this model is that the vibration of each tube is not affected by the vibration of adjacent tubes and the fluid is regarded flowing in a channel between the heat exchanger. The schematic diagram of the flow tube model is as follows.

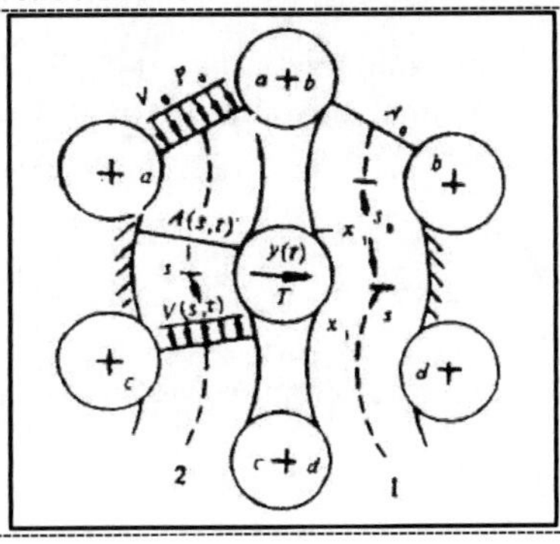

Figure 1. Schematic diagram of flow tube model.

According to the research, the vibration direction of the tube is mainly perpendicular to the flow direction. The tube transverse does simple harmonic vibration, and the displacement equation can be expressed as:

$$y(t) = y_0 sin\omega t \tag{1}$$

$$A_i(s,t) = A_0 + (-1)^{i+1} A(s)\sin\left[\omega t + \psi(s)\right] \quad i{=}1,2 \tag{2}$$

$$U_i(s,t) = U_0 + (-1)^{i+1} U(s)\sin\left[\omega t + \varphi(s)\right] \quad i{=}1,2 \tag{3}$$

$$P_i(s,t) = P_0 + (-1)^{i+1} P(s)\sin\left[\omega t + \theta(s)\right] \quad i{=}1,2 \tag{4}$$

Among the equations, $A_i(s,t)$ represents the area of the unsteady flow tube. A_0 represents the unit length steady flow tube area at the entrance. $A(s)$ indicates the disturbance quantity of unit length flow tube area. $\psi(s)$ is the flow tube area phase difference. $U_i(s,t)$ indicates an unsteady flow rate. U_0 represents the inlet flow rate. $U(s)$ is the velocity disturbance quantity. $\varphi(s)$ is the phase difference of the velocity. $P_i(s,t)$ represents unsteady pressure. P_0 is the entrance pressure. $P(s)$ indicates pressure disturbance quantity. $\theta(s)$ indicates the pressure difference for the pressure. In those equations, the first one is the steady state term and the second one is the disturbance term.

The continuity equation of one-dimensional unstable incompressible fluid can be expressed as:

$$\frac{\partial A_i(s,t)}{\partial t} + \frac{\partial A_i(s,t)U_i(s,t)}{\partial s} = 0 \tag{5}$$

Derivation of equation (2) is

$$\frac{\partial A_i(s,t)}{\partial t} = (-1)^{i+1} y_0\omega\cos\left[\omega t + \frac{2s\omega}{U_0}\right] \tag{6}$$

Make equation (5) and (6) joint order, and integral along s:

$$(-1)^{i+1} y_0\omega \int_{-s_0}^{s} \cos\left[\omega t + \frac{2s\omega}{U_0}\right] ds + A_i(s,t)U_i(s,t)|_{-s_0}^{s} = 0 \tag{7}$$

Substituting equation (3) and (4) into equation (7):

$$U(s)=\frac{3U_0y_0}{A_0}\sin\frac{s+s_0}{U_0}\omega \tag{8}$$

$$\varphi(s)=-\frac{\pi}{2}+\frac{s-s_0}{U_0}\omega \tag{9}$$

The velocity distribution of the flow field can be determined by the former equation.
The differential form of the one-dimensional unstable incompressible fluid energy equation can be expressed as:

$$\frac{d}{dt}\left(\rho\frac{U_i^2}{2}\right)=\rho\left[\frac{\partial}{\partial t}\left(\frac{U_i^2}{2}\right)+U_i\frac{\partial}{\partial s}\left(\frac{U_i^2}{2}\right)\right]=-U_i\frac{\partial P_i}{\partial s} \tag{10}$$

After sorting out the equations, and integrating along s, we can obtain:

$$\begin{aligned}P(s)\sin\left[\omega t+\theta(s)\right]&=\frac{3U_0^2\rho y_0}{4A_0}\sin\omega t\left[3\cos\left(\frac{2\omega s}{U_0}\right)-3\cos\left(\frac{2\omega s_0}{U_0}\right)-\frac{2\omega(s+s_0)}{U_0}\sin\left(\frac{2\omega s_0}{U_0}\right)\right]\\&+\frac{3U_0^2\rho y_0}{4A_0}\cos\omega t\left[3\sin\left(\frac{2\omega s}{U_0}\right)+3\sin\left(\frac{2\omega s_0}{U_0}\right)-\frac{2\omega(s+s_0)}{U_0}\cos\left(\frac{2\omega s_0}{U_0}\right)\right]\end{aligned} \tag{11}$$

According to the pressure distribution on the both sides of the flow tube, the excited force can be obtained, and the equation can be expressed as:

$$F_p(t)=l\int_{-s_1}^{s_2}\left[P_1(s,t)+P_2(s,t)\right]ds \tag{12}$$

Among them, the stable phase of P_1, P_2 is uniform, the wave phase has the same size, but opposite direction, then:

$$F_p(t)=F_1\sin\omega t+F_2\cos\omega t \tag{13}$$

The vibration equation of the tube can be expressed as:

$$F_p(t)=ml\frac{\partial^2y}{\partial t^2}+C\frac{\partial y}{\partial t}+ky \tag{14}$$

$$C=\frac{ml\omega\delta_0}{\pi} \tag{15}$$

m indicates the quality per unit length of the heat transfer tube. k is the elastic coefficient. C represents the damping coefficient. δ_0 is the logarithmic decay rate.
The vibration equation can be expressed as:

$$ml\frac{\partial^2y}{\partial t^2}+\left(C-\frac{F_2}{\omega y_0}\right)\frac{\partial y}{\partial t}+\left(k-\frac{F_1}{y_0}\right)y=0 \tag{16}$$

Heat transfer tubes vibrate under the action of fluid elastic excitation. The vibration equation is:

$$\ddot{x}_j^n+2\delta_0 f\dot{x}_j^n+4\pi^2f^2x_j^n=\frac{F_j^n}{m_0} \tag{17}$$

In this equation, x represents the displacement of the tube vibration. f represents the natural frequency of the tube. F indicates the lift and drag force of the tube under fluid action. m is the quality of the tube. j=1, 2.
The displacement equation of the tube produced by the force of fluid elastic excitation is:

$$\ddot{x}_j^n = \frac{F_j^n - kx_j^n}{m_0} - 2\delta f \dot{x}_j^n \tag{18}$$

$$\dot{x}_j^{n+1} = \dot{x}_j^n + \ddot{x}_j^n \times \Delta t \tag{19}$$

$$x_j^{n+1} = x_j^n + \frac{1}{2}\left(\dot{x}_j^n + \dot{x}_j^{n+1}\right)\Delta t + \frac{1}{2}\ddot{x}_j^n \times \Delta t^2 \tag{20}$$

2.2. Computational model

There are many boundary deformation and motion problems in engineering simulation. The dynamic mesh model in FLUENT software can be used to simulate the flow problem of calculation area change caused by boundary deformation or motion [15]. The feature of dynamic mesh model is that mesh can be reconstructed according to the needs of the different processes in the solution. FLUENT software can be used to observe the function relationship between the transverse displacement and the time history to reflect the strength of elastic excitation of the heat transfer tubes in the flow field.

Use Pro / Engineer software to create the model, and then initial meshing. In order to analyse vibration of the tube at different positions, we select three regions as objects of the analysis of numerical simulation: the shell side of the entrance, the shell side of the central and shell side of the exit. In order to be closer to the actual situation, the nine adjacent tubes of each region are also taken into account (figure 2). The finite element model is figure 3.

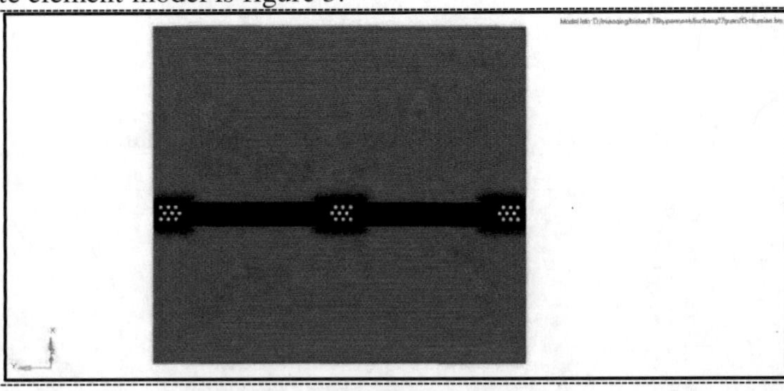

Figure 2. Finite element model for the numerical simulation of the flow around a cylinder in a heat exchanger.

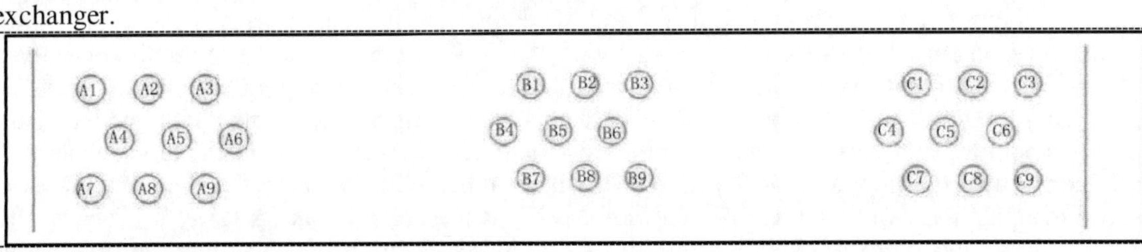

Figure 3. Schematic diagram of the position number of the heat exchange tubes.

2.3. The boundary conditions of the simulation

The finite element model was imported to the FLUENT software. The basic solver is set to the pressure based, transient, implicit, absolute and so on. Selecting the RNG k- ε model as calculation model. Set the material parameters for the medium, the entry and exit boundary conditions, velocity inlet. And input the incoming fluid speed. Select pressure outlet, input static pressure. Set shell side, tube board surface, partition surface and the surface of heat transfer tube as wall. The wall is a non-slip. And activate the dynamic mesh model, select the mesh update model as the spring smoothing model and local region reconstruction; deploy dynamic mesh fields, assume that the movement of the rigid body uses UDF to define the movement of the boundary, the motion equation of the heat transfer tube under the fluid excitation force obtained above defines the movement of the boundary of the heat transfer tube. Set the time step to 0.00001s, 1000 times, iterate 20 times at every turn.

3. Results and discussion

Obtain velocity distribution after simulating (figure 4).

Figure 4. Velocity distribution of the heat exchanger.

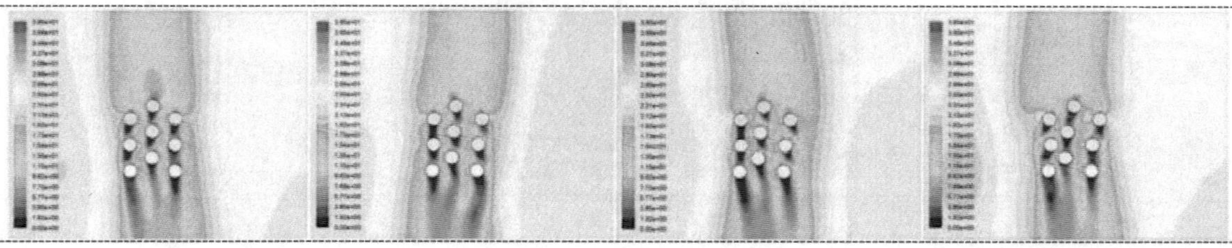

Figure 5. Schematic diagram of vortex shedding in heat exchanger's shell side.

The nine tubes in the central area was taken to observe the velocity distribution as preliminary analysis. Founding that the boundary of the heat transfer tubes is constantly changing and vortex shedding phenomenon obviously. This proves that it is correct that using the dynamic mesh model in the solution condition. It can also be found that the flow field near the outer edge of the tube is changing more intense than the flow field at the edge of the inner tube. With the increase of time, the area of wake is more obvious. But the area of wake is not large. If the area of wake is large, it is shown that the heat transfer tube is more likely to produce fluid induced vibration, and the vibration first caused by the vortex shedding and turbulence buffeting. These vibrations cause partial displacement of the heat transfer tube bundles. And then with the continuous increase of flow velocity, fluid elastic instability is likely to induce heat transfer tube bundles vibration. As the outer flow field changes more intense, the outer tube bundles displacement changes more obvious than the inner tube bundles displacement, and outer tube bundles are more likely to produce flow induced vibration.

We found that the locations of different tubes are constantly changing, and the changing trend is different. Such as figure 5, the right side of a row of three heat exchanger tubes. The connection of these three heat exchanger tubes is nearly a straight line at the first picture, but it becomes a curve at the fourth picture. And based on the second and the third picture, it is unable to analyse the variation tendency and law of relative position of heat exchanger tubes. Therefore, it is necessary to use FLUENT to monitor and record the displacement of each heat exchanger tube boundary separately.

After monitoring the displacement of all the heat exchanger tubes by FLUENT, The variation of the lateral displacement of 27 heat exchangers with time is obtained. Comparing with the heat exchanger tubes at different locations in different areas, we get the following results as figure 6 and figure7.
According to figure 6, the displacement of the heat exchanger tubes at the same position is quite different at different region, the heat exchanger tubes displacement of region A changed with time is much larger than B and C, the maximum displacement reached 3 units during the monitoring period (one unit is 0.001mm). It shows that the intensity of the heat exchange tube in region A is the strongest. It also shows that at position 4, the heat exchanger tubes vibration is less significant at A, B, C three regions, the displacements are all within 1 unit. It means that in different areas, position 4 is not easy to produce a strong vibration. figure 7 shows that region A is more prone vibration and the vibration in the position 4, position 5, and position 6 in region A are less intense. The vibration in position 5 and position 6 in region B is relatively stronger, which is the primary vibration location in the region B. Only position 6 vibration is more significant in the region C.
In general, the heat exchanger tubes vibration of region A (the shell side of the entrance) is relatively stronger when the heat exchanger tubes at the same position in different region. And heat transfer tubes vibration in different position in the same region show different characteristics.

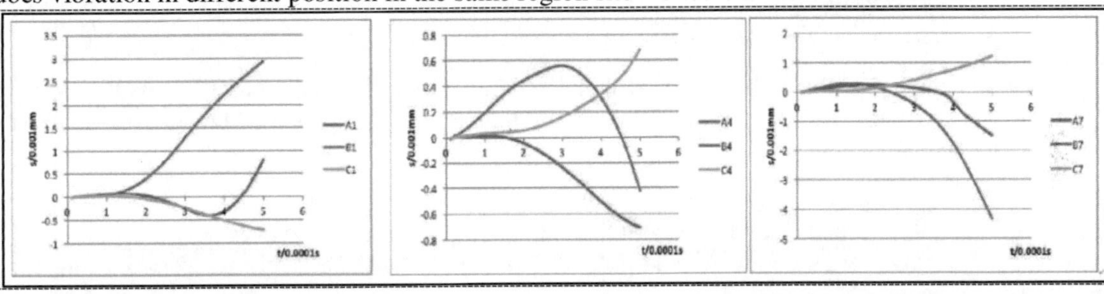

Figure 6. Displacement time curve of tube at same locations in different areas.

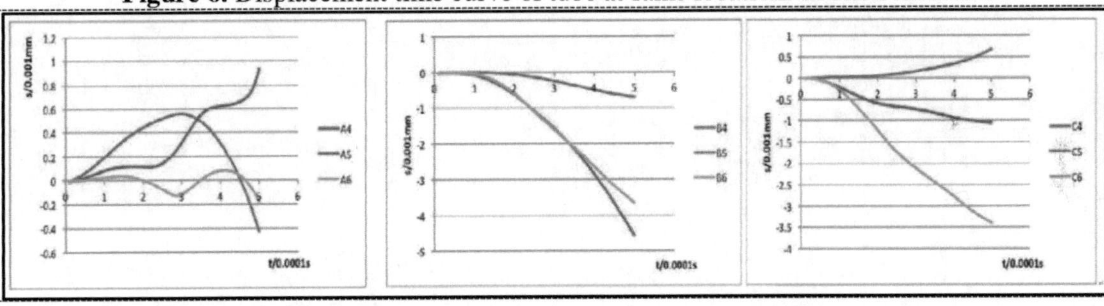

Figure 7. Displacement time curve of tube at different locations in same areas.

4. Conclusions

The background is the failure of tube bundles in air heat exchanger in our paper. We use the FLUENT software dynamic grid model to simulate the flowing status of the flow field around the cylindrical group. Kaman vortex can be seen and the fluid flow is observed to cause the cylinder moving (vibration response). We control the variable, comparing with the different parts of the heat exchanger tube at the same location on the displacement. It shows that the shell side of the tube generated by the vibration displacement is greater than the other location of the tube. So the tubes in the shell side of the entrance are susceptible to failure due to fluid induced vibration.

5. References

[1] Qian S W 2002 *Handbook for Heat Exchanger Design* (Beijing: China Chemical Industry Press)
[2] Costa A L H and Queiroz E M 2008 *Appl Therm Eng* **28**
[3] Fesanghary M, Damangir E and Soleimani I 2009 *Appl Therm Eng* **29** 5-6
[4] Ravagnani M A S S, Silva A P, Biscaia E C and Caballero J A 2009 *Ind Eng Chem Res* **4**

[5] Feng G 2012 *Chem Ind Eng Prog* **31**, 3
[6] Su Y C 2012 *A Study on the Flow-Induced Vibration and Heat Transfer Characteristic of Elastic Tube Bundle* (Shandong: Shandong University)
[7] Qian S W 1988 *Flow Induced Vibration of Heat exchanger Foundation* (Wuhan: Huazhong College of Technology Press)
[8] Jiang C Y 2012 *Study of Calculation Method of Flow Induced Vibration in Tube and Shell Heat Exchanger* (Shanghai: Fluid machinery and Engineering, East China University of Science and Technology)
[9] Li A J, Xing G J and Zhou L W 2007 *Liaoning Chemical Industry* **36**
[10] Bo H L 1994 *J Tsinghua Univ: Nat Sci Ed* **34** 6
[11] Bo H L 1995 *Chin J Nucl Sci Eng* **15** 1
[12] Nie Q D 1996 *Acta Mech Sinica* **28** 2
[13] Lever J H, Weaver D S 1982 *J. Pressure Vessel Technol* **104** 3
[14] Lever J H, Weaver D S 1986 *J Sound Vib* **107** 3
[15] Zhu H J 2011 *Fluent 12 Fluid Analysis and Engineering Simulation* (Beijing: Tsinghua University Press)

Acknowledgments
This work was supported by "The Fundamental Research Funds for the Central Universities" (No. DUT15QY35) and "Dalian University of Technology Innovative Undertaking Management Platform" (No. 2016101410837).

IOP Conf. Series: Materials Science and Engineering **239** (2017) 012012 doi:10.1088/1757-899X/239/1/012012

Influence of Rolling Force and Roll Gap on Thickness of Strip

Wenyu MA, Fengqin WANG, Xibang ZHANG, Wei GUO, Ziying LIU, Lin TAN

Beijing Key Laboratory of Green Recycling Process for Iron & Steel Production Technology,
Shougang Research Institute of Technology, Beijing, 100043, China

Wenyu MA (wymaustb@163.com)

Abstract. The automation control system is of great significance and efficient to control the thickness precision in hot strip rolling process. In this paper, the relationship among thickness, rolling force and roll gap in hot strip rolling process was studied. The reason for the bad hit rate of thickness for the first format-changed blank was analyzed. The adjustment of thickness in the hot rolling process was also studied. The decrease of rolling force and the increase of roll gap can increase the exit thickness of the hot-rolled strip. As the thickness control system and mechanism for hot rolling process are complicated, there are much work need to be done.

1. Introduction

In 1924, a hot strip rolling line of 1470 mm was established in America, from then the hot strip rolling technology has a long history of more than 90 years [1]. In November 1971, Nippon Steel established a 2235 mm hot strip rolling with computer controlled system, which is a typical advanced design and technology [2]. In 1978, Wuhan steel company has import the first 1700 mm hot rolling strip production line with strip hot rolling computer control system [3]. The application of process automation system significantly improve the efficiency of the strip hot rolling and is advanced in the modern computer control system. At the same time, using the process automation system can improve the thickness control precision and ensure the product quality [4-5]. Thickness precision is one of the most important quality indexes in the area of hot strip. Much effort has been devoted to control the strip thickness precisely. And the method to control the thickness is an important research in the rolling automation [5-7]. According to a hot rolling line in Shougang company, much attention has been paid to the accuracy of thickness control, such as theoretical research, optimization development, data analysis and other parts. In order to control the thickness precisely with the help of computer control system, some work has been done, such as collecting and summarizing thickness related data form the log, pdo, pda and other profiles generated from level 2 computer system, studying the theory of thickness variation of hot rolled strip, analyzing the relationship among thickness and other factors and so on.

2. Materials

In this rolling schedule, the steel 64A02 was used. The blank of 64A02 was usually used for building, kitchen utensils, tableware, vehicles, household appliances and so on. In our industry, the hot rolling strip was produced. The entry thickness in first finish rolling stand is about 45 mm, and the exit thickness of strip in the last finish rolling stand is about 3.5 mm, which is the target thickness for our customer. However, the thickness mentioned above should consider the effect of temperature. At the elevated temperature, the thickness of 3.5 mm can be converted to a hot-stated value of about 3.54, by multiply the coefficient of thermal expansion. The coefficient of thermal expansion is about 1.01 at that elevated temperature.

3. Experiments

The transfer bar was transferred to finish rolling stands. There are 6 stands in the hot rolling stage. In this stage, more than 10 blanks was rolled. After the hot rolling process, the thickness was measured by X-ray measurement equipment. At the same time, the rolling force of each stand was also measured and recorded in the computer. Then we can get these data from the log, pdo, pda and other profiles for further analysis. As we all know, the accuracy of thickness is of great significance, and there are so many problems need to be explained and solved. If the thickness accuracy can be accurately controlled or improved, the sale can be enhanced. So it is necessary to analyze the data about the thickness, rolling force, roll gap and the relationship among them.

Table 1 shows the thickness related data for these three blanks. The type of steel is 64A02, the target width is 1198 mm, the target thickness is 3.5 mm, and the average thicknesses for these three strips are 3.301, 3.504 and 3.502 respectively. The hit rates of thickness for these three strips are 0.26, 92.5 and 78.5 respectively. The final rolling temperatures for these three strips are 876.1, 881.4, and 879.2 respectively. The inherence modes for these strips are Long, short and short respectively. As the first one has a different size from the last one, so the hit rate of thickness is low. The second one learned from the first one, and got a better hit rate of thickness. As is shown in Table 1, the hit rate of thickness for the first blank is 0.26, which is a little low. And the second one has a high hit rate of 92.5. This is to say the computer system adjusted the coefficient of the model, and some data were changed, so the average thickness was improved to 3.504 mm. The third one also has a good average value of 3.502.

In order to analyze the reason for the variation of thickness. Some other two important factors, namely rolling force and set roll gap, were selected and given in Table 2 and 3.

Table 1. The information for the first three blank

N o.	Steel grade	Target width	Target thickness	Average thickness	Hit rate	Rolled temperature	Inherenc e mode
1	64AO2	1198.	3.50	3.301	0.26	876.1	LONG
2	64AO2	1198.	3.50	3.504	92.50	881.4	SHORT
3	64AO2	1198.	3.50	3.502	78.50	879.2	SHORT

Table 2. The rolling force for the first three blanks.

Rolling force of head	1	Calculated [MN]	33.131	28.75	24.759	21.447	16.58	14.065
		Actual [MN]	27.557	23.68	21.463	18.592	14.462	12.502
		Deviation (Cal-Act)	5.574	5.07	3.296	2.855	2.118	1.563
	2	Calculated [MN]	25.561	23.93	19.576	17.518	13.907	11.752
		Actual [MN]	25.345	23.739	19.889	16.635	13.444	11.345
		Deviation (Cal-Act)	0.216	0.191	-0.313	0.883	0.463	0.407
	3	Calculated [MN]	26.287	23.541	21.661	17.36	12.702	11.033
		Actual [MN]	25.812	23.573	21.056	17.236	12.69	11.311
		Deviation (Cal-Act)	0.475	-0.032	0.605	0.124	0.012	-0.278

Table 3. The set roll gap for these three blanks.

			F1	F2	F3	F4	F5	F6
Roll gap for	1	Set roll gap [mm]	22.596	13.137	7.227	4.47	2.622	2.418
the stands	2	Set roll gap [mm]	23.791	13.282	7.872	5.04	3.001	2.814
	3	Set roll gap [mm]	23.678	13.69	7.561	4.878	3.169	2.924

4. Analysis and discussion

It can be seen from Table 1 that the inherence mode of the first blank is Long term inherence. Because this blank changed the format from the last one. Even though the steel grade is the same as the last one 64A02, the change of format led to the long term inherence. However the inherence result is not good, even shows a large variation from the reality. However, the second one and the third one show the good result, and the inherence modes are short term inherence respectively. After the check for the inherence. It can be known that the second one learned from the first one, and the third one learned from the second one. The first one learned from the NeuralNet. This should be why the first blank has a bad thickness distribution.

The thickness of the hot rolled slab can be analyzed by P-H diagram, as shown in Fig. 1. In this figure, the left two lines represent stand modules, and the right two lines represent material modules. The h indicates the target exit thickness. This figure is a schematic diagram used for further analysis.

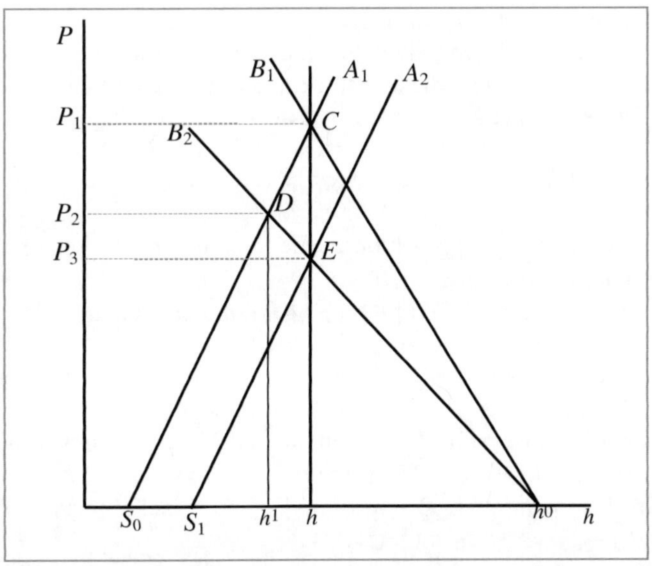

Figure 1. P-H schematic diagram for thickness variation analysis

The thickness is mainly controlled by rolling force and roll gap. It can be seen from Table 2 the deviation between calculated and actual result of rolling force for the first one is very large. The deviation for the second one and third one is small. At the same time, the calculated result of rolling force for the second one is less than the first one. The set roll gap for the second one becomes larger from the first one. The adjustment of rolling force and roll gap generated a good result for the second slab. According to the P-H diagram, it can be concluded the material modules for the first slab were set as the false ones. After the first roll process, the control system detected this error, and changed the material modules, so the second one and third one have the good thickness distribution. In the P-H diagram, the line B1C was drawn as the first material modules, it has a large slope, which indicates a large material module. The line B2D was drawn as the new material modules after adjustment of computer system. It has a relatively small slope, which indicates a small material modules. In order to check the material modules set by the

IOP Conf. Series: Materials Science and Engineering **239** (2017) 012012 doi:10.1088/1757-899X/239/1/012012

Level-2 computer system, the stand and material modules were output into the log profile. The data were summarized in Table 4. It can be seen that the precise material modulus set is of great significant for the thickness control in hot rolling process. At the same time, the rolling force and roll gap have great effect on the slab thickness. The decrease of rolling force and the increase of roll gap can increase the exit thickness of the slab.

Table 4. The stand modules and material modules obtained from computer control system

Sample	Modules	F1	F2	F3	F4	F5	F6
1st	Stand module [MN/mm]	6.158	5.933	6.291	5.734	5.532	5.556
	Material mod [MN/mm]	1.723	2.917	3.714	7.505	9.565	17.073
2nd	Stand module [MN/mm]	6.194	5.953	6.306	5.649	5.551	5.534
	Material mod [MN/mm]	1.33	2.253	3.247	6.145	8.265	15.531
3rd	Stand module [MN/mm]	6.194	5.951	6.284	5.648	5.483	5.522
	Material mod [MN/mm]	1.368	2.27	3.416	5.733	8.599	15.298

5. Summary and Conclusions

In this paper, the data about thickness for the hot rolling slap were presented and analyzed. The slap firstly changed format tends to have a bad thickness distribution, due to the inherence study. The application of material modules affect the exit thickness of slap significantly. The rolling force and roll gap have a remarked influence on the thickness. The decrease of rolling force and the increase of roll gap would increase the thickness of slab.

6. References

[1] Wenxue Zhang. Research and application of the thickness control strategy of hot strip mill [D], Shenyang: Northeast University, 2010.

[2] Tao Sun. Research and application of high precision thickness control for medium and heavy plate [D], Shenyang: Northeastern University, 2009.

[3] Bo Huang. Technology of hot strip mill process and equipment [J], Iron and Steel, 2002, 37: 474-476.

[4] Wen Peng. Process Control System Development and Model Optimization for Hot Strip Finishing Mill [D], Shenyang: Northeastern University, 2014.

[5] Zhang G M, Xiao H, Wang C H. Three-dimensional Model for Strip Hot Rolling[J], Journal of Iron and Steel Research, 2006, 13(1): 23-26.

[6] MEI Rui-Bin, LI Chang-sheng, LIU Xiang-hua, et al. Analysis of Strip Temperature in Hot Rolling Process by Finite Element Method[J], Journal of Iron and Steel Research, international, 2010, 17(2), 17.

[7] Yang Y Y, Linkens D A, Talamantes-silva J, et al. Roll Force and Torque Prediction Using Neural Network and Finite Element Modelling[J], ISIJ International, 2003, 43(12): 1957-1966.

7. Acknowledgement

All the authors would like to thank Lijie Dong, Guang Yang from Shougang Qian' an Iron & Steel Company very much for their kind guideline and help.

IOP Conf. Series: Materials Science and Engineering **239** (2017) 012013 doi:10.1088/1757-899X/239/1/012013

Flux Growth of Highly Crystalline Photocatalytic $BaTiO_3$ Particle Layers on Porous Titanium Sponge Substrate and Insights into the Formation Mechanism

Q Wang[1] and B Li[1*]

[1] School of Mechanical and Power Engineering, East China University of Science and Technology, 130 Meilong Road, Shanghai 200237, P. R. China

[*] Corresponding author. Tel: +86-21-64252601; E-mail: bingli@ecust.edu.cn.

Abstract. A unique architecture of idiomorphic and highly crystalline $BaTiO_3$ particle layers directly grown on a porous titanium sponge substrate was successfully achieved for the first time using a facile molten salt method at a relatively low temperature of 700 ℃. Specifically, the low-melting KCl-NaCl eutectic salts and barium hydroxide octahydrate were employed as the reaction medium and barium source, respectively. Powder X-ray diffraction (XRD), scanning electron microscopy (SEM), Energy dispersive X-ray spectroscopy (EDS), transmission electron microscopy (TEM), high-resolution transmission electron microscopy (HRTEM) and UV-vis diffuse reflectance spectrophotometry were used to characterize the structure, morphology and optical property of the obtained samples. The results revealed that the flux-grown tetragonal $BaTiO_3$ products had well-defined and uniform morphology with an average size of 300 nm and a band gap of ~3.16 eV. Based on XRD, EDS, SEM, and TEM, the possible formation mechanism responsible for the well-developed architecture of $BaTiO_3$ particle layers was proposed and discussed. Furthermore, the photocatalytic activity of the flux-grown $BaTiO_3$ products for organic pollutant degradation under simulated sunlight irradiation was also investigated.

1. Introduction

Semiconductor photocatalysis has attracted extensive attention due to its potential applications in photodegradation of organic pollutants, H_2 generation from water splitting, CO_2 reduction into hydrocarbon fuels and the removal of heavy metal ions, which greatly contributes to the solution of worldwide problems of energy crisis and environmental pollution[1,2,3]. It is well-known that the development of highly efficient, cost-effective and stable photocatalysts is crucial for the further advancement and large-scale practical applications of photocatalytic technique. During the past few decades, tremendous efforts have been devoted to developing new-type photocatalytic materials, including metal oxides (TiO_2, ZnO, etc.), metal sulfides (CdS, MoS_2, etc.), titanates ($SrTiO_3$, $BaTiO_3$, etc.), vanadates (Ag_3VO_4, $BiVO_4$, etc.), tungstates ($ZnWO_4$, $CdWO_4$, etc.), metal phosphates (Ag_3PO_4, $BiPO_4$, etc.), niobates ($NaNbO_3$, $Bi_5Nb_3O_{15}$, etc.), metal-free semiconductor (g-C_3N_4), and the related composite photocatalysts[4,5,6].

However, most photocatalysts still suffer from the shortcomings such as high recommbination rate of photogenerated electron-hole pairs, small specific surface area and low sunlight utilization efficiency. Moreover, the separation and recovery of nanophotocatalysts from the treated effluent should be also concerned for practical industial application[7,8]. Therefore, developing novel and highly efficient photocatalysts with rational structural design is still a challenging task for the applications of photocatalysis in solar energy utilization and environmental remediation.

In this work, a unique architecture of idiomorphic and highly crystalline $BaTiO_3$ particle layers directly grown on porous Ti sponge substrate was prepared for the first time by the facile molten salt method in KCl-NaCl flux at 700 ℃. On one hand, the porous structure of Ti sponge is beneficial to improve the adsorption properties of the $BaTiO_3$ products and then facilitate the photocatalytic reactions. Moreover, it is reported that the band bending nature of tetragonal $BaTiO_3$ ferroelectric material can inhibit the recombination of charge carriers, thereby contributing to the enhancement of its photocatalytic performance[8,9,10]. On the other hand, the flux-grown $BaTiO_3$ photocatalysts attached on the Ti sponge can be easily reclaimed from the environment without secondary pollution. Based on the characterization results including XRD, SEM, EDS and TEM, the corresponding formation mechanism of flux-grown $BaTiO_3$ particle layers on porous Ti sponge can be demonstrated clearly under the controlled experimental conditions, which can provide a new approach to fabrication of other highly efficient photocatalysts for the practical energy and environmental applications.

2. Experimental

In a typical procedure, the porous Ti sponge materials were ultrasonically cleaned in dilute hydrochloric acid, deionized water, acetone and ethanol in sequence and dried in air. Afterwards, Ti sponge and $Ba(OH)_2\cdot 8H_2O$ were ground homogenously with NaCl-KCl eutectic salts (50 mol% NaCl + 50 mol% KCl) at a molar ratio of 1: 1: 20, and then transferred into a corundum crucible sealed with a lid. Subsequently, the mixture was calcined at 700℃ in an electric furnace, and then cooled naturally to room temperature. The products were repeatedly washed with warm deionized water to remove residual salts, and finally dried at 90℃ overnight. According to the holding time (1 h, 4 h, 8 h and 12 h), the obtained samples were denoted as BTO-1, BTO-2, BTO-3 and BTO-4, respectively.

The flux-grown samples were systematically characterized by XRD (Rigaku D/MAX 2550 VB/PC), SEM (Hitachi S-3400N), EDS (EDAX, Genesis XM2), TEM/HRTEM (JEOL JEM-2100), UV-vis diffuse reflectance spectrophotometry (Varian Cary 500 spectrophotometer). The photocatalytic activity of the obtained samples was evaluated by photodegradation of a 50 mL aqueous methyl orange (MO, 10 mgL^{-1}) solution in the presence of 50 mg sample under simulated sunlight irradiation using a 300 W Xe lamp (λ>300nm, PLS-SXE300) at room temperature. Prior to light irradiation, the suspension was magnetically stirred in dark until absorption-desorption equilibrium. The change of characteristic absorption of MO solution at λ= 465 nm corresponding to its concentration was monitored on the Varian Cary 500 spectrophotometer to evaluate the photocatalytic activity.

3. Results and Discussion

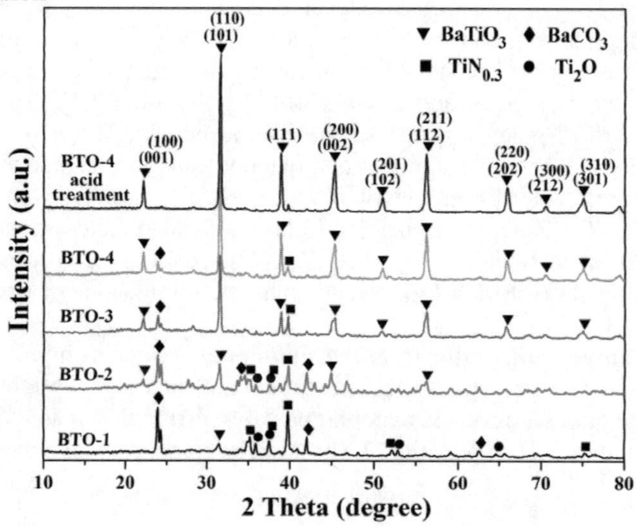

Figure 1. XRD patterns of the as-synthesized samples.

As shown in Figure 1, XRD patterns of the obtained samples reveal that BTO-1 is mainly composed of $BaCO_3$ (JCPDS Card NO. 05-0378), $TiN_{0.3}$ (JCPDS Card NO. 41-1352), Ti_2O (JCPDS Card NO. 65-5293), and tetragonal $BaTiO_3$ (JCPDS Card NO. 05-0626), but the amount of obtained $BaTiO_3$ phase is relatively low, which can be attributed to the primary process of the formation of Ti-rich $TiN_{0.3}$ and Ti_2O layers on Ti substrate during limited calcination time. Obviously, with the increasing of calcination time, the amount of $BaCO_3$, $TiN_{0.3}$ and Ti_2O decrease gradually, while the amount of $BaTiO_3$ increase obviously. Meanwhile, the enhancement of diffraction peaks due to $BaTiO_3$ can suggest the increase in the degree of grain growth and crystallinity of the flux-grown $BaTiO_3$ crystals. When the calcination time reached 12 h, a large amount of tetragonal $BaTiO_3$ phase was obtained, although it coexisted with certain unreacted $BaCO_3$ (which could be easily removed by an acid treatment) and $TiN_{0.3}$. Moreover, the sharp and intense peaks indicate the flux-grown $BaTiO_3$ products are highly crystalline.

Figure 2. Typical SEM images and EDS spectra (inset) of the samples: (a) Ti sponge (b, c) BTO-1, (d, e) BTO-2, (f) BTO-3, (g, h) BTO-4 and (i) TEM (inset, HRTEM) image of BTO-4.

Based on the XRD results, the SEM and TEM images in Figure 2 display clearly the formation process and morphology evalution of the flux-grown crystalline $BaTiO_3$ particle layers on Ti sponge substrate. Figure 2a shows the Ti sponge has a porous structure with smooth surface. As shown in Figure 2b-c combined with the corresponding EDS analysis, the surface of BTO-1 is covered by compact Ti-rich Ti_2O nanoparticles (out layer) with an averge size of ~100 nm because no N element can be detected, and thus the inner layer consists of Ti-rich $TiN_{0.3}$ phase. Here, no obvious $BaTiO_3$ crystals can be found owing to their low amount. Figure 2d reveals that BTO-2 is composed of a large amount of amorphous and aggregated particles with the size less than 200 nm, while the $BaTiO_3$ products can hardly be distinguished from the mixtures. Interestingly, as can be seen in Figure 2e, the cross-sectional SEM image shows that the inner layer of BTO-2 exhibits wormlike morphology, suggesting a remarkable structural evalution of inner layer in comparison with that of BTO-1. As for BTO-3, well-crystallized $BaTiO_3$ particles with a larger particle size can be seen in Figure 3f, which is in accordance with the XRD results. As shown in Figure 2g-h, BTO-4 has a unique architecture of well-defined $BaTiO_3$ polyhedral crystal layers densely grown on Ti substrate. The small and armorphous particles can be assigned to the residual $BaCO_3$. In Figure 2i, TEM and HRTEM images reveal that the flux-grown $BaTiO_3$ polyhedral crystals with the size of 100-300 nm are highly crystalline because no obvious defects can be observed in clear lattice fringes.

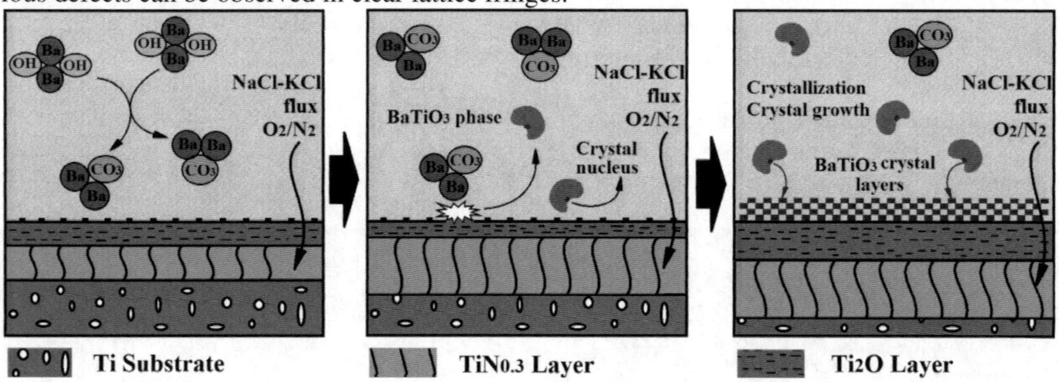

Figure 3. Schematic illustration of formation mechanism responsible for the well-developed $BaTiO_3$ crystal layers grown on porous titanium sponge substrate.

Based on the above results and our previous studies[11,12], the formation mechanism responsible for the flux-grown $BaTiO_3$ crystal layers on porous Ti sponge was proposed, as illustrated in Figure 3. The films of Ti-rich Ti_2O (out layer) and $TiN_{0.3}$ (inner layer) were firstly formed on the surfaces of Ti sponge by the nitridation and oxidation reactions (i.e., $2Ti+0.5O_2 \rightarrow Ti_2O$, $Ti+0.15N_2 \rightarrow TiN_{0.3}$) at the solid-liquid interface between Ti sponge and NaCl-KCl flux. Simultaneously, $Ba(OH)_2$ was transformed into $BaCO_3$ by reacting with CO_2 from the ambient. Generally speaking, the flux growth of well-defined polyhedral crystals can be ascribed to the dissolution and recrystallization process. With the increasing of calcination time, the $BaTiO_3$ phase generated by the reaction (namely, $Ti_2O + 1.5O_2 + 2BaCO_3 \rightarrow 2BaTiO_3 + 2CO_2\uparrow$) began to be dissolved in the NaCl-KCl flux and then $BaTiO_3$ crystallites (i.e., crystal nucleus, serving as templates for further growth) would form and grow in-situ on Ti_2O layer with the increasing of solute concentration (i.e., $BaTiO_3$ phase). Meanwhile, the $TiN_{0.3}$ was transformed into Ti_2O by reacting with atmospheric oxygen, which can further continually promote the generation of $BaTiO_3$ phase. Therefore, according to the dissolution-precipitation mechanism, the precipitation of new-formed $BaTiO_3$ on the aforementioned templates continued until the reactants were consumed completely, leading to the crystal growth and shape evaluation of the $BaTiO_3$ crystals.

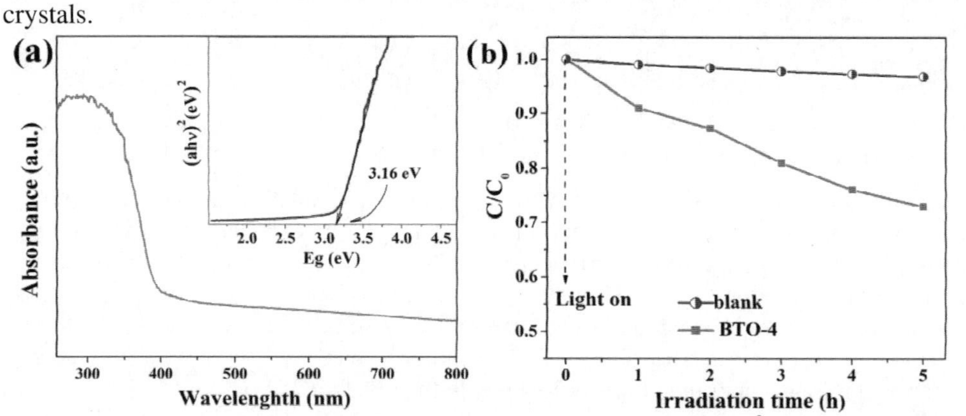

Figure 4. (a) UV-vis spectra of the sample BTO-4. Inset is the plot of $(\alpha h\nu)^2$ versus $h\nu$ for calculating the Eg. (b) Photocatalytic degradation of MO over the flux-grown $BaTiO_3$ products.

As shown in the UV-vis diffuse reflectance spectra (in Figure 4a), BTO-4 mainly absorb the UV light below about 400nm. The band gap energy (Eg) of sample BTO-4 was calculated to be ~3.16 eV, according to the equation: $\alpha h\nu = A(h\nu - Eg)^{1/2}$, where α, *hv*, and A are the absorption coefficient, photon energy and a constant, respectively, which aggrees with the previous reports[8,13]. Figure 4b displays the photocatalytic degradation of MO over the flux-grown $BaTiO_3$ products, revealing that the photocatalytic activity needs to be further improved to meet the requirements of practical environmental and energy applications, which is similar to the previous studies[8, 9, 13]. This can be mainly attributed to the high recombination rate of photogenerated electron-hole pairs and the weak visible-light harvesting ability of $BaTiO_3$. Considering that the construction of semiconductor heterostructures has been intensely studied as an effective way to improve the photocatalytic activity, coupling the $BaTiO_3$ particles with other visible-light-driven photocatalysts (such as g-C_3N_4, MoS_2, Ag_3VO_4, etc.) to enhance their photocatalytic properties will be carried out.

4. Conclusion

In summary, a novel architecture of idiomorphic $BaTiO_3$ polyhedral crystal layers directly grown on a porous Ti sponge was successfully developed in KCl-NaCl flux at 700 ℃. The flux-grown tetragonal $BaTiO_3$ products with well-defined shape and uniform morphology had a Eg of ~ 3.16 eV. The possible formation mechanism responsible for the well-formed $BaTiO_3$ particle layers was proposed based on the corresponding structural and morphological characterization. Finally, the simulated sunlight photocatalytic activity of the obtained $BaTiO_3$ products for MO degradation was also tested and discussed. Moreover, our work on further enhancing the photocatalytic performance of flux-grown $BaTiO_3$ photocatalyst is underway. Overall, this work offers a new strategy for designing other easily recycled photocatalytic materials with rational architectures favorable for practical application in solar energy conversion and environmental remediation.

5. References

[1] Hernández-Alonso M D, Fresno F, Suárez S and Coronado J M 2009 *Energy Environ. Sci.* **2** 1231
[2] Tu W, Zhou Y and Zou Z 2014 *Adv. Mater.* **26** 4607
[3] Li H, Zhou Y, Tu W, Ye J, Zou Z 2015 *Adv. Funct. Mater.* **25** 998
[4] Zhao Z, Sun Y and Dong F 2015 *Nanoscale* **7** 15
[5] Wang Q, Guo Q, Wang L and Li B 2016 *Dalton Trans.* **45** 17748
[6] Wang H, Zhang L, Chen Z, Hu J, Li S, Wang Z, Liu J and Wang X 2014 *Chem. Soc. Rev.* **43** 5234
[7] Akhundi A and Habibi-Yangjeh A 2016 *Mater. Chem. Phy.* **174** 59
[8] Fan H, Li H, Liu B, Lu Y, Xie T and Wang D 2012 *ACS Appl. Mater. Interfaces* **4** 4853
[9] Cui Y, Briscoe J and Dunn S 2013 Chem. Mater. **25** 4215
[10] Kappadan S, Gebreab T W, Thomas S and Kalarikkal N 2016 *Mater. Sci. Semicon. Proc* **51** 42
[11] Wang Q, Guo Q, Hu Y and Li B 2016 *CrystEngComm* **18** 6926
[12] Wang Q, Guo Q and Li B 2015 *RSC Adv.* **5** 66086
[13] Xian T, Yang H, Di L J and Dai J F 2015 *J Alloys Compd.* **622** 1098

Acknowledgments
This work was supported by National Natural Science Foundation of China (No.51274102).

Tribological Behaviour under Conformal and Non-Conformal Contact Condition of Piston Ring and Cylinder Liner Material in a Reciprocating Bench Test

H. K. Trivedi [1, a], **Dr D. V. Bhatt** [2]

[1]Research Scholar and [2]Professor, SVNIT, Surat, India

[a]hetalktrivedi@gmail.com

Abstract. Experiments on cylinder liner and piston ring were carried out on a reciprocating tester to measure the coefficient of friction. The tribological behaviour for conformal and non-conformal contacts was compared. Actual cylinder liner is and piston ring is used for conformal contact, which is cut as per the required dimension. After analyzing the composition of the liner material, a specimen is prepared for non-conformal contact. Tests were conducted for variable load, variable speed and variable temperature with various commercial lubricants to measure the friction coefficient. The result shows that conformal contact shows significant variation in friction coefficient as compared to non-conformal contact for all operating parameters. The geometry of contact, oil quantity, and viscosity of lubricant plays a vital role to characterize the behaviour of friction coefficient and wear.

Keywords: Friction, Wear, Cylinder liner, Piston ring, conformal and non-conformal contact.

1. Introduction

The engine is a major responsible component in vehicular design in the petrol and diesel engines where a great deal of energy loss at the interface of cylinder liner wall surface and piston ring[1]. The factors which are responsible for friction losses are gas pressure forces due to combustion, ring tension, hydrodynamic pressure, inertia force due to mass of the ring, speed and temperature of engine and asperity contact forces [2]. Due to the mechanical motion engine oil is used as a lubricant to reduce friction and wear. The properties of the lubricants used in the engine plays an important role in to extend the life and improve the performance of an engine [3]. The operating conditions and performance of the contact also depend on the shape of the bodies in contact[4].

The physic-chemical and tribological properties of commercial engine oil plays a vital role and it is observed that the coefficient of friction and load have an inverse correlation [5]. Keller [6] has understood the topography on the tribology cast iron and observes that friction coefficient mainly influenced by surface asperities and an initial presence of these asperities produces the high friction coefficient. Sutaria [7] developed an algorithm for calculation of the cylinder pressure and piston ring assembly friction with crank angle under unfired single cylinder four-stroke gasoline engines. The results indicate that piston speed, piston ring tension and viscosity of lube oil are playing an important role in the friction forces. Riyadh [8] has performed an experiment on a test rig and observed that coefficient of friction decreases as the load increases due to the temperature rise between the two sliding surfaces and its effect on the viscosity of lubrication. Truhan [9] has investigated that coefficient of friction is dependent on the load and temperature. Mixed or boundary lubrication condition exists at room temperature. Peggy [10] has conducted experiments and analyzes the effect of low phosphorous engine oil that after surface break-in all the oil containing friction modifiers observed reduces a coefficient of friction. To understand the interactions between the surface tension, the surface energy of the solid phase and their influence on the coefficient of friction, it is observed that wth the increasing oil temperature, the stribeck curve minimum is shifted to a higher sliding speed due to lower viscosity [11]. In this work, tribological behaviour for conformal and non-conformal contact for cylinder liner and piston ring were compared. Various commercial lubricants were used for the experimental work to measure the friction coefficient.

2. Experimental setup

2.1 Specimen preparation

The material used for liner-specimen was taken from the market. Test samples were cut from real engine parts to keep original curved surface finishing. For a non-conformal contact, specimens were prepared as per the dimension of test rig holder. After analyzing the composition and microstructure of liner material, cast iron plates were prepared by casting and subsequently turned and grounded. The hardness of the plates was 79.8HRC and the surface roughness was 0.365±0.2μm. The piston ring segments were taken from the market which is actually used in Honda engine with a capacity of 100CC. Piston ring was cut into two halves to fix on the groove of the ring attachment holder. The surface roughness of piston ring was 2.75±0.05 μm.

2.2 Experimental Apparatus and Procedure

Tests were carried out with a reciprocating friction and wear monitor test rig (TR-281M-M6). A piston ring segment reciprocates against the counter cylinder liner segment. In this work, tests were performed with a range of load, speed, temperature and lubricants for conformal and non-conformal contact. Prior to each test, the ring and liner were cleaned by acetone to remove metal fragments and oil from the surface. Before start, the test, 2drops of oil is added to the contact surface of liner and piston ring. New specimens were used for each test. The Win Ducom software was used for data acquisition and display of results. An electronic weighing balance (accuracy of 0.0001mg) was used to measure the weight loss of the specimens. The above procedure is repeated for all tests.

Figure 1. Reciprocating test rig

Figure 2. Holder and liner with conformal contact

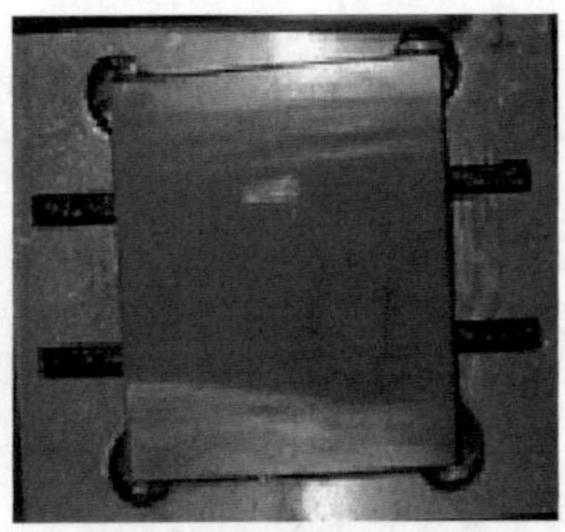

Figure 3 Holder and liner with non – conformal contact

3. Results and Discussion

A series of experiments were carried out for conformal and non-conformal contacts to measure the friction coefficient and he results are graphically illustrated

Table 1 Oil Properties

Lubricants	Viscosity(cSt)		Flash Point (°C min)	Viscosity Index
	40°C	100°C		
SAE10W30	119.75	11.51	228	102
SAE20W40	124	14.3	240	114
SAE20W50	156.3	17.87	256	126

Table 2 Material composition of prepared cylinder liner on X-met 5000-portable Analyzer

Elements	Fe	C	Si	Mn	Cr	Cu	P	S
Weight (%)	Rest	3.16	2.05	0.67	0.3	0.27	0.21	0.062

3.1 Frictional behaviour

3.1.1 Effect of speed

Figure 4 and figure 5 summarizes the effect of speed for various lubricants on the friction behaviour of sliding pair of the cylinder liner and piston ring for conformal and non-conformal contacts. With the increase of speed, for conformal contact, the friction coefficient range from 0.178 to 0.04

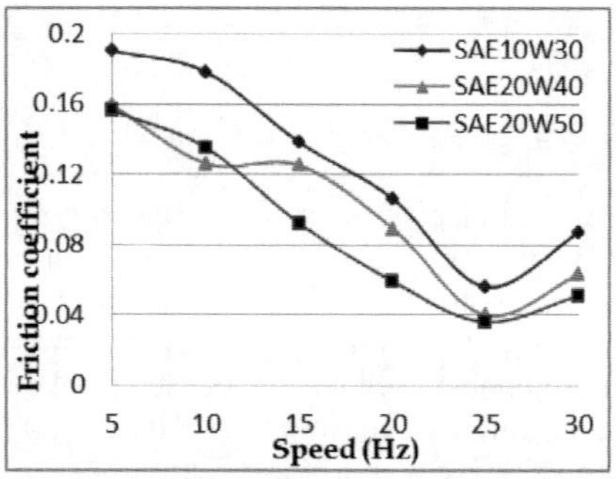

Figure 4. Effect of speed on friction coefficient for CC

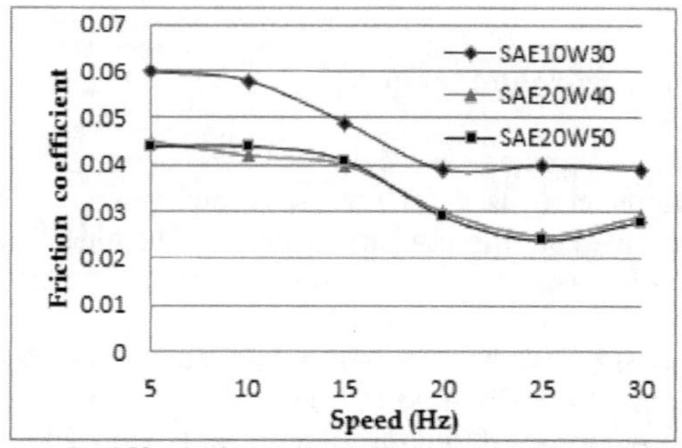

Figure 5. Effect of speed on friction coefficient for NCC

while for non-conformal contact, the range is 0.06 to 0.03. It shows that contact area and the quantity of lubricant play a vital role. From the graph, it is observed that at lower operating speed, it offers highest friction coefficient. At lower operating speed, a hydrodynamic action is not strong, so the film thickness may be small [12]. Smaller film thickness leads to an increase in the asperity contact between the contact surfaces, results in high friction coefficient. For non-conformal contact, elastohydrodynamic influence is more evident [13]. This effect is visible for all sliding speeds. With the further increase of speed, the hydrodynamic pressure developed between the contact surfaces and shearing between the contact surfaces becomes smaller. This may be due to the mixed film developed between the contact surfaces. Afterwards, a large amount of hydrodynamic pressure generated and negligible shear can occur between the surfaces. Due to that effect, contact surfaces completely separated by the fluid film and the friction is generated due to the shearing of the lubricant fluid film which is dependent on the rheological properties of the lubricant. The result is friction coefficient again increases.

3.1.2 Effect of load

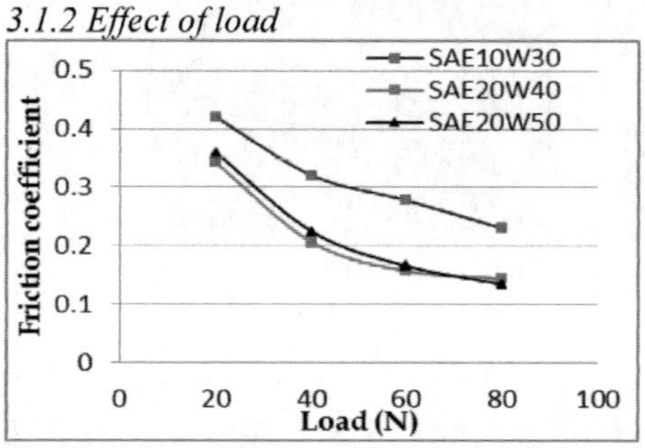

Fig.6. Effect of load on friction coefficient for CC

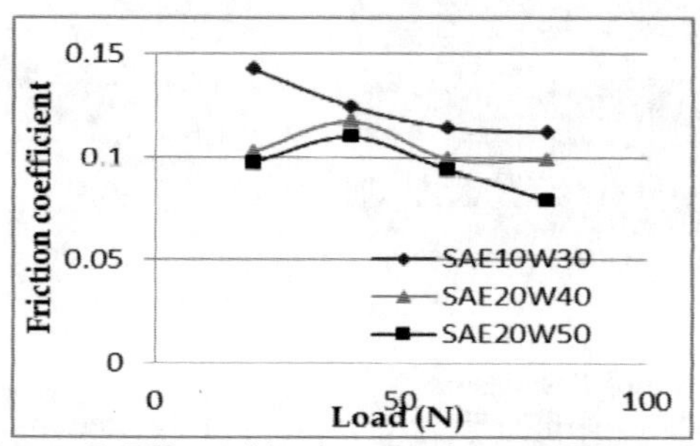

Figure 7. Effect of load on friction coefficient for NCC

Above figure shows the effect of load on the coefficient of friction at 600 rpm and ambient temperature for various lubricants. The nature of curve for all lubricants is similar and follows a negative trend for conformal contact while for non-conformal contact only little reduction takes place. For conformal contact, the friction coefficient range from 0.420 to 0.135 while for non-conformal contact, the range is 0.143 to 0.079. During the initial stage, surface irregularities play a major role for friction coefficient. With the effect of small load, no more contact area generated between the specimens. Boundary lubrication regime is developed between the surfaces which may be responsible for increasing the friction coefficient. With the increase of load, surface irregularities in the form of asperities increase the contact area between the sliding surfaces. The surface becomes smooth due to the high wear rate hence mixed lubrication is developed. Lubricant formed film which may be stable due to the smooth geometry between surfaces. In non-conformal contact, full film lubrication condition exists between the mating surfaces which are responsible to developed elastohydrodynamic lubrication. The result is little reduction in friction coefficient.

3.1.3 Effect of temperature

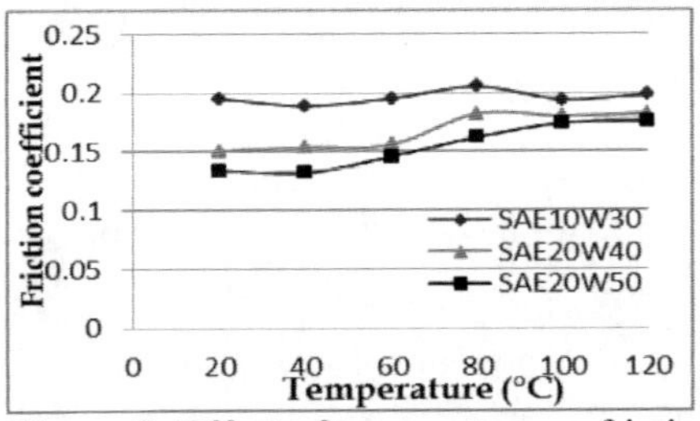

Figure 8. Effect of temperature on friction coefficient for CC

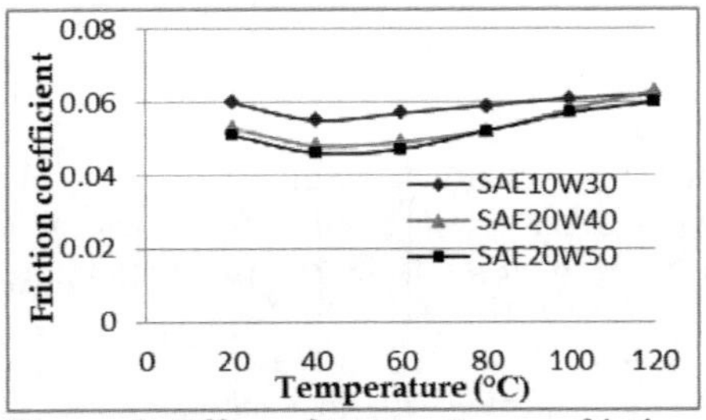

Figure 9. Effect of temperature on friction coefficient for NCC

It is observed from the figure that the friction coefficient changes with change of temperature for conformal and non-conformal contact both. SAE10W30 offers stable performance compare with SAE20W40 and SAE20W50. This may be due to SAE20W40 and SAE20W50 are a more viscous lubricant, so they might be more affected with temperature as compare to SAE10W30. With the increase of temperature, the viscosity of the lubricant reduced and it became thinner so it generates asperity contact between the mating surfaces. This change the lubrication regime from mixed to boundary which may be responsible for increasing the friction coefficient.

From the experiments, it was found that conformal contact shows more friction coefficient as compare to non-conformal contact. In non-conformal contact, the graph shows almost stable or little change in their performance. Here, the same amount of oil is added for both types of contacts. The contact area covered by the specimen is quite more in conformal contact which offers high friction coefficient. This may be due to the large covered area; lubricant is pushed aside during its one stroke and does not flow back in time for the return stroke. It causes starvation which reduced oil film thickness due to the inadequate volume of the lubricant present at inlet [14] [15].

With non-conformal contact, sufficient amount of lubricant is placed between the liner and piston ring. The full film can generate which reduce the metal to metal contact and moves under elastohydrodynamic regime result are lower friction coefficient. It means that friction and wear are more affected with conformal contact only. In a real engine, there are conformal contacts generating between the cylinder liner and piston rings. So to analyse the wear behaviour, conformal contact is taken for the further experiments.

3.2 Weight loss of cylinder liner and piston ring

The experiment is performed for a constant load 80N and duration is 100 min to calculate the weight loss of the cylinder liner and piston ring for various lubricants.

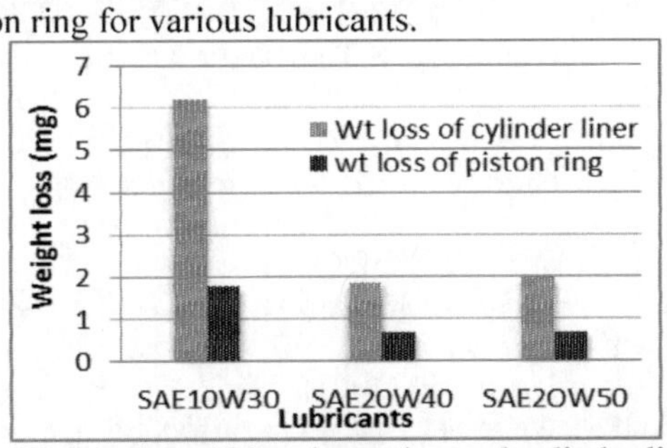

Figure 10. Weight loss of the piston ring and cylinder liner for CC

Figure 10 shows the comparisons of the weight loss of cylinder liner and piston ring. Cylinder liner offers more weight loss as compare to piston ring and SAE10W30 have a major weight reduction as compare to SAE20W40 and SAE20W50. The cylinder liner shows 2 to 5times greater weight loss as compare to the piston ring. This relationship is reversed in an actual operating engine [16, 17].

Here, cylinder liner offers 70.96%, 61.08% and 65% higher weight loss as compare to the piston ring. The wear loss with SAE10W30 lubricant for piston ring is 43.64% and 34.55% and for cylinder liner is 63.80 and 62.86% high as compare to cylinder liner and piston ring. Viscosity plays a vital role in this phenomenon. SAE10W30 is a low viscous lubricant so it may be not capable to form a fluid film between the sliding surfaces result is more wear occurs.

In an actual engine, the piston ring experiences more wear as compare to cylinder liner but at laboratory scale, it's vice versa. In a combustion chamber, variations in pressure, temperature and gas composition occurs which are difficult to reproduce in a laboratory test [17]. There are dissimilarities in geometry between the ring and liner and instead of cycling loading, it perform constant load throughout the operation which results high wear generates in a test rig [18].

4. Conclusion

Based on the presented experimental outcomes, following conclusions have been drawn.

- Conformal contact offers high friction coefficient for all operating condition. In this case, the same amount of oil is added for both contacts but the contact area covered by the conformal contact is quite more. So starvation occurs due to that thin film formed which caused boundary regime and increases friction coefficient. The operating conditions and performance of the contact are strongly depends on the contact geometry of the specimen.
- Mixed lubrication plays the significant role in decreasing the friction coefficient for conformal contact while elastohydrodynamic influence is more evident with non-conformal contact.
- SAE10W30 offers higher friction coefficient as compare to SAE20W40 and SAE2W50 lubricants for all operating parameters. It shows that the quality of the lubricants plays a vital role on tribological behaviour.
- There are dissimilarities in geometry and working condition between the test ring and actual engine. So during the wear analysis of the specimen, the cylinder liner shows 2 to 5times greater weight loss as compare to the piston ring. This relationship is reversed in an actual operating engine.

References

[1] J. Heywood: *Int Combustion Engine Fund*, McGraw-Hill, USA (1988), p 730.
[2] K. Wannatong, S. Chanchaona, and S. Sanitjai: *Simulation Modelling Pract and Theory*, **16** (2008) 127.
[3] C. M. Taylor: *Wear* 221 (1998), 1.
[4] M. Torbacke and S. K. Rudolphi: *Lubricants: Introduction to properties and performance*, Wiley (1994), p 115.
[5] G. D. Thakre and M. R. Tyagi: *Ind Lubri Trib*, **65** (2013) 5.
[6] J. Keller, V. Fridrici, P. Kapsa and J. F. Huard: *Trib I J*, **42** (2009), 6.
[7] B. M. Sutaria, D. V. Bhatt and K. N. Mistry: *Proc WCE*, (2009) p 1452.
[8] A. Riyadh, Haftirman, R. A. Khairel, Y. Al-dourt: *J Surface Engg mat Adv Tech*, **2** (2012), 3.
[9] J. J. Truhan, J. Qu and P. J. Blau: Trib Int J, **38** (2005), 3.
[10] L. Peggy, G. Barber, Q. Zoo, J. R. Anderson, S.Tong and A.Quintana: *Trib Tran*, **51** (2008), 5.
[11] Mathias, K. Norbert, H. Falk and B. Niklas: *Mat Performane Characterisation*, 3 (2014) 1.
[12] M. Cater, N. W. Bolander and F. Sadeghi: *SAE Technical Paper,* **32** (2006) 4.
[13] Hamrock, Schmid & Jacoson: *Fundamental of fluid film lubrication*, Marcel Dekker Inc, New York (2004), p 49.
[14] S. C. Vladescu and T. Reddhoff: *Wear*, 358 (2016).
[15] Ma, I. Sherrington, E.H. Smith: *Proc. Inst. Mech. Eng. Part J – J Engg Trib*, **210** (1996), 3.
[16] B. Kim, H. C. Young and Heung: *Trib Int* **70** (2014).
[17] J. J. Truhan, J. Qu and P. J. Blau: *Wear,* **259** (2005), 7.
[18] M. G. Naylor, P. Kodali and J. Wang: *Diesel engine tribology - Modern Tribology Handbook*, CRC Press LLC, (2001) p. 1231.

IOP Conf. Series: Materials Science and Engineering **239** (2017) 012015 doi:10.1088/1757-899X/239/1/012015

Effect of Atmosphere on Recovery Dynamics of Polyimide Film Damaged by Electron Radiation

E. Plis[1], D. P. Engelhart[2], R. Cooper[3], S. Humagain [4,6], M. Brunetti[5,6], A. Koch[7], S. Greenbaum[4], D. Ferguson[3], R. Hoffmann[3]

[1] Assurance Technology Corporation, USA
[2] National Research Council at Air Force Research Laboratory, USA
[3] Air Force Research Laboratory, Space Vehicles Directorate, USA, Kirtland AFB, NM, USA
[4] Department of Physics & Astronomy, Hunter College of the City University of New York, USA
[5] Physics Department, New York City College of Technology, The City University of New York, Brooklyn, NY 11201, USA
[6] The Graduate School and University Centre, the City University of New York, New York, NY 10016, USA
[7] Aerospace Engineering Department, Worcester Polytechnic Institute, Worcester, MA 01606, USA
plis@assurtech.com

Abstract. Since electrons are the primary charged particles at the geosynchronous Earth orbit (GEO), understanding of their interactions with spacecraft materials, such as polyimide (PI, Kapton-H®), is important. Understanding of the chemical nature of electron damage and its effect on PI's electrical and optical properties is still limited. Thus, predictive spacecraft models (electrical charging, thermal, etc) are restricted to only pristine material properties. This is a major source of error in spacecraft construction and anomaly resolution, since material properties change after exposure to the space environment. Ground based measurements are critical to understanding the dynamics of spacecraft materials however it will be shown in this work that standard material handling practice and exposure to air are unacceptable for these studies.

1. Introduction

Nowadays, polyimide (PI) films are widely utilized in the spacecraft industry for their insulating properties, mechanical durability, low density, and resistance to radiation and atomic oxygen damage [1]. In particular, Kapton-H ®, a solid organic insulator with a monomeric unit of $C_{22}H_{10}N_2O_5$, is often used as an ultra-lightweight thermal control blanketing, also known as multi-layer insulation or MLI, to provide stable range of operating temperatures for satellite systems [2].

Spacecraft materials are subjected to the harsh effects of space weather, including hard vacuum, atomic oxygen, ultraviolet light, charged particle radiation, etc. causing degradation of their optical and physical properties [3]. It is important to understand the alterations of spacecraft materials properties' induced by the space environment to improve satellite lifetime, system reliability, space situational awareness, and decrease satellite operational costs. Whereas change of PIs' material properties under simulated space environments, including surface degradation, mechanical properties, conductivity, and absorptivity, has attracted attention of researcher, relation of these radiation-induced changes to the chemical nature of damage still needs to be investigated [4-7].

In general, it has been found that the space environment fundamentally changes spacecraft materials, in particular, their surface becomes very reactive due to the breakage of inter-molecular bonds and formation of free radicals as a result of space environment. Ground based spacecraft material degradation, or aging, typically takes place under ultra-high vacuum conditions, ~ 10^{-9}-10^{-11} Torr, the subsequent material storage and characterization is often performed under uncontrolled laboratory atmosphere. Residual gases, even at low partial pressures, can react with radicals in the modified PIs, changing the properties of aged material in a way that is not properly accounted for. A small number of publications have investigated the effect of humidity on the electrical behavior of undamaged polyimide [8, 9]. However, to the authors' knowledge, little attention has been given to the effects that the storage atmosphere has on the material properties of irradiated PI films.

In the presented study, we first characterized the material properties of electron-irradiated PI as a function of exposure to uncontrolled laboratory atmosphere (air). Second, we evaluated the effects of several major constituents of Earth's atmosphere (Ar, N_2, O_2) on the material properties of electron irradiated PI.

2. Experimental

In terms of total energetic dose experienced by an object in geosynchronous Earth orbit (GEO) over several solar cycles, electrons are the dominant source of energy deposition. In order to simulate an electron dose typical for GEO environment, 76.2 μm thick Kapton-H® films were irradiated with high energy (90 keV) mono-energetic electron radiation gun [10]. Irradiation was performed at an accelerated dose rate of 94.8 Gy/s with vacuum level maintained at lower than 1 x 10^{-6} Torr for the entire duration of the radiation exposure. PI samples in this study were irradiated with doses of 2.3 x 10^7 Gy and 5.5 x 10^7 Gy, equivalent to 9 years and 17 years of GEO electron exposure, respectively.

The conductivity of PI samples exposed to air after the irradiation process was measured using the constant voltage method [11]. Conductivity of irradiated PI samples recovered in vacuum under partial pressure of nitrogen, oxygen, or argon gases was derived from a measurement of the surface potential decay [12]. After the PI samples were irradiated with high energy electrons, the radiation-damaged PI film was removed from the aging facility and transferred to the characterization vacuum chamber or portable vacuum container for transfer to the EPR measurement facility. Sample transfer was performed under laboratory atmosphere (air) and was limited to less than 5 min of air exposure. The radical spin population was obtained by comparing the integrated intensity of PI sample spectra with that for a standard sample a known concentration of 4-hydroxy-2,2,6,6-tetramethylpiperidin-1-oxyl (TEMPO).

3. Results and Discussion

Interaction of PI with highly energetic particles will modify its chemical structure. The extent of this modification is a function of several simultaneous kinetic processes, namely, damage (interaction of material with highly energetic particles, resulting in broken chemical bonds), healing (formation of bonds identical to those damaged, returning the material to its pristine state), and scarring (formation of new chemical bonds in the damaged material which are different from those in the pristine material). Often macroscopic properties are measured, making it difficult to distinguish healing from scarring; hence, we refer to the sum of healing and scarring as recovery. Radiation-induced chemical modifications will induce changes in PI's physical properties, in particular, bulk conductivity [13].

Fig. 1 compares bulk conductivities of Kapton-H® samples irradiated with dose of 5.6 x 10^7 Gy and recovered in air (top panel) and under vacuum (bottom panel). Initial conductivity of damaged PI, measured at time zero, was nearly the same for air-recovered (5 x 10^{-15} $(\Omega\cdot m)^{-1}$) and vacuum-recovered (8 x 10^{-15} $(\Omega\cdot m)^{-1}$) samples, suggesting that our estimation of bulk conductivity of aged PI is reliable. Air exposure of radiation-damaged PI resulted in fast recovery, within 3 hours, of irradiated material's conductivity from 5 x 10^{-15} $(\Omega \cdot m)^{-1}$ to the one of the pristine Kapton (2 x 10^{-18} $(\Omega \cdot m)^{-1}$), whereas the vacuum recovery of radiation-damaged PI retained the same value, ~ 8 x 10^{-15} $(\Omega \cdot m)^{-1}$, for over three weeks (504 hours) after the damaging process.

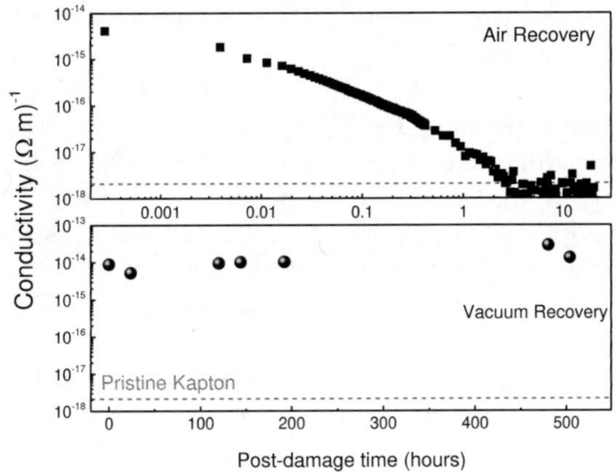

Figure 1. Comparison of air- and vacuum-recovered conductivities of Kapton-H® irradiated with dose equivalent to 2.3 x 10^7 Gy Dashed line indicates the conductivity of pristine Kapton-H®.

To further investigate an effect of damaging radiation on PI, including the concentration and nature of free radicals as well as mechanism of their decay, irradiated PI was sealed in a vacuum EPR tube for EPR measurements. The double integral of the derivative line shape of the EPR spectra was calculated to derive their corresponding radical concentrations. Fig. 2 demonstrates the evolution of radicals concentration in PI film irradiated with dose of 5.5 x 10^7 Gy and recovered in air and under vacuum.

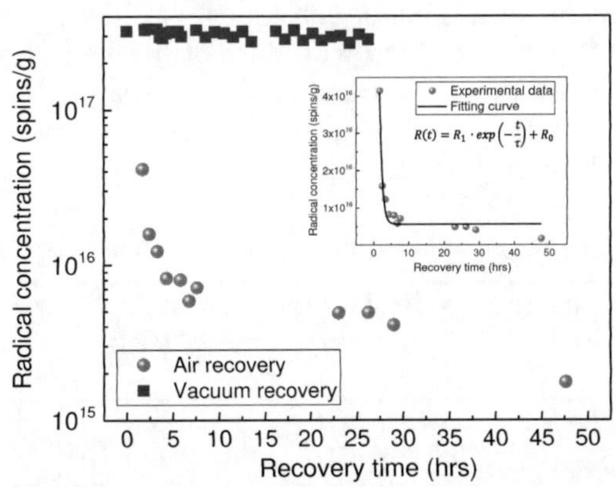

Figure 2. Radical concentration of PI irradiated with dose of 5.5 x 10^7 Gy after air- and vacuum-recovery. Inset shows theoretical fitting to the air-recovery data as well as fitting equation. See text for more details.

To further investigate an effect of damaging radiation on PI, including the concentration and nature of free radicals as well as mechanism of their decay, irradiated PI was sealed in a vacuum EPR tube for EPR measurements. The double integral of the derivative line shape of the EPR spectra was calculated to derive their corresponding radical concentrations. Fig. 2 demonstrates the evolution of radicals concentration in PI film irradiated with dose of 5.5 x 10^7 Gy and recovered in air and under vacuum.

Under vacuum, the concentration of radicals remains constant in the irradiated PI sample, ~ 3 x 10^{17} spins/g, whereas a decay in the radical concentration is observed in air-exposed irradiatied PI material. The first EPR measurement of air-exposed irradiated PI was taken after 1.7 hours of ambient exposure, and radical concentration was lowered by an order of magnitude compared to the radical concentration during the vacuum recovery.

The decay of radicals in the irradiated polymer can proceed via two mechanisms, thermal recombination of the radicals and reaction with the gases in the ambient atmosphere. The time evolution of the radical concentration under air exposure may be fitted using the rate equation shown in the Fig. 2 inset. Here R(t) is the total radical population, spins/g; R_1 and R_0 are constants representing the annealed free-radical population and the surviving radical population under the measurement condition (in ambient at room temperature), respectively; τ is the characteristic time constant which is related to the recovery rate of the annealable radicals. The resulting fit is presented in Fig. 2 inset.

As may be seen from inset to Fig. 2, the rate equation describes well the evolution of the radical population in irradiated PI under air exposure until ~ 8 hours of air exposure, and beyond that it fails to fit the experimental data. This observation is consistent with results reported by Wu *et al* [14]. In [14], authors attributed the fast time decay phase of radical evolution to the thermodinamical recombintation of radicals, whereas the slow decay regime is attributed to the reaction of radicals with the gases in the ambient atmosphere. It should be noted that even the in damaged material that is exposed to air the radical consentration does not return to the prisitine value of zero.

Although the kinetic model of Wu *et al* describes our data well, the results shown in Fig 2 show conclusively that the initial rapid decay of the radical concentration cannot be explained with thermodynamic model. If the rapid decay of the radical signal were thermodynamic, the EPR signal would decay as quickly in vacuum as in air, since in vacuum the PI sample is in thermal equilibrium with the surrounding chamber and sample mount, i.e. 300 K. However, the radical signal measured in vacuum is shown to be stable for > 30 hours. The rapid decay in radical concentration in atmosphere can therefore only be due to the presence of external reactants, such as oxygen or water vapor.

To further evaluate the effects of several major constituents of Earth's atmosphere on the material properties of electron irradiated PI, PI irradiated with dose of 5.5 x 10^7 Gy was stored in a vacuum chamber at a background pressure of 1×10^{-6} Torr and exposed to partial pressures of argon, nitrogen, and oxygen gases. Surface potential decay (SPD) measurements were performed after each individual gas exposure to determine bulk electrical conductivity [31].

Table1. Details and results of the "environmental" experiments.

Environment	Exposure (L)	Resistivity (Ω·m)	Conductivity $(\Omega\cdot m)^{-1}$	Exposure time (hrs)
Vacuum	N/A	5.1 x 10^{15}	1.9 x 10^{-16}	N/A
Argon	35 x 10^{12}	7.8 x 10^{15}	1.3 x 10^{-16}	74.5
Nitrogen	90 x 10^{12}	4.8 x 10^{15}	2.1 x 10^{-16}	125.0
Oxygen	14 x 10^{12}	3.2 x 10^{17}	3.1 x 10^{-18}	92.0

As shown in Table 1, the transport properties of irradiated PI did not change significantly with exposure to an argon or nitrogen environment. However, exposure to an oxygen atmosphere resulted in a degradation of two orders of magnitude in the electrical conductivity compared to that of vacuum-exposed film, from 1.9 x 10^{-16} $(\Omega\cdot m)^{-1}$ to 3.1 x 10^{-18} $(\Omega\cdot m)^{-1}$. This may be attributed to the reaction of radicals in irradiated PI with oxygen molecules. This reaction transforms the radicals into the new chemical species with no unpaired electrons, thus reducing the number of low-lying trap states in the irradiated material and degrading its conductivity.

4. Summary

Electrical transport properties of radiation-damaged and recovered (both in vacuum and ambient atmosphere) PI films were investigated via EPR and conductivity measurements. We found that even limited exposure to the uncontrolled laboratory atmosphere dramatically affects the properties of interest. In particular, the conductivity of irradiated PI decreased by an order of magnitude, from 5 x 10^{-15} $(\Omega\cdot m)^{-1}$ to 5 x 10^{-16} $(\Omega\cdot m)^{-1}$ after only 2 minutes of air exposure. Concentration of radiation-induced radicals measured by EPR was decreased by an order of magnitude after 1.7 hours of air exposure, from 3 x 10^{17} spins/g (initial value) to 4 x 10^{16} spins/g, respectively. These results are consistent with those we observed earlier [15].

Further, we evaluated the effects of several major constituents of Earth's atmosphere, nitrogen, oxygen, and argon, on the material properties of electron irradiated PI. We found that argon and nitrogen atmosphere do not effect the conductivity of irradiated PI, whereas oxygen exposure results in two orders of magnitude degradation of electrical conductivity of irradiated PI from 1.9 x 10^{-16} $(\Omega\cdot m)^{-1}$ to 3.1 x 10^{-18} $(\Omega\cdot m)^{-1}$.

Thus, we have shown that while PI is an exceptionally stable polymer, damaging radiation makes it sensitive to its surrounding atmosphere. The damaged material's sensitivity to even limited air exposure means that the storage, handling, and characterization protocols of irradiated material must be scrutinized and carefully controlled. Argon or nitrogen atmosphere may be suitable candidates for the assessement of radiation-damaged PI's material properties, resulting in development of reliable models guiding the design of novel technology-enabling materials.

Acknowledgments
Authors would like to acknowledge support from AFOSR grants 17RVCOR401 and 17RVCOR414. We would also like to thank Dr. John Weaver for the assistance performing EPR measurements.

References

[1] Y. Kimoto, T. Fujita, N. Furuta, A. Kitamura, H. Suzuki. J Spacecr Rockets 1-6 (2016)
[2] H. A. MacEwen, G. G. Fazio, M. Lystrup, N. Batalha, N. Siegler, E. C. Tong. Proc, of SPIE 9904, 990405 (2016)
[3] R. K. Kiefer, W. J. Gabler, M. T. Hovery, S. A. Thiibeault. Rad Phys Chem 80 (2), 126 (2011)
[4] X. Lei, M. Qiao, L. Tian, Y. Chen, Q. Zhang. Corros Sci 98, 560 (2015)
[5] H. Shimamura and T. Nakamura. Polym Degrad Stab 95 (1), 21 (2010)
[6] R. Hanna, T. Paulmier, P. Molinie, M. Belhaj, B. Dirassen, D. Payan, N. Balcon. Appl Phys 115, 033713 (2014)
[7] A.K. Sharma, N. Sridhara. Adv Space Rese 50(10): 1411 (2012)
[8] F. Belluci, I. Khai,s. S. D. Senturia, R. M. Latanision. J Electrochem Soc 137, 1778 (1990)
[9] N. A. Galichin, M. E. Borisova. Russ Electr Eng 78, 129-132 (2007)
[10] R. Cooper and R. Hoffmann, Air Force Report No. AFRL-RV-PS-TP-2015-0012 (2015)
[11] ASTM-D257-14, "Standard Test Methods for DC Resistance or Conductance of Insulating Materials," 2015
[12] J. R. Dennison, J. Brunson, P. Swaminathan, N. W. Green, A. R. Frederickson. IEEE Trans Plasma Sci 34, 2191 (2006).
[13] E. Plis D. P. Engelhart, R. Cooper, and R. Hoffmann. Phys. Stat. Sol B (2017)
[14] Y. Wu, C. Sun, J. Xiao, R. Li, D. Yang, S. He. Polym Degr Stab 95,1219 (2008)
[15] D. P. Engelhart, E. Plis, D. Ferguson, R. Cooper, R. Hoffmann. Presentation at AMOS 2016 conference.

IOP Conf. Series: Materials Science and Engineering **239** (2017) 012016 doi:10.1088/1757-899X/239/1/012016

Research on the Unsteady Discharge Flow of Dry Chemical Powder Tank

HeeGeun Song, Sedong Kim, Kitae Yu, Cheol Park, YoungGil Jung, Hyomin Jeong (Corresponding)

Department of Energy and Mechanical Engineering, Gyeongsang National University
Cheondaegukchi-Gil 38, Tongyeong, Gyeongnam, 650-160, South Korea
gmlrms1016@gmail.com

Corresponding email: hmjeong@gnu.ac.kr

Abstract. In this study, we investigated the activation of powders by the supply of N2 gas in the dry chemical powder tanks. In this study, we investigated the effect of N2 gas on the activation of dry chemical powder(KHCO3) in Dry chemical Powder Tank, We studied how dry chemical powder(KHCO3) is activated according to the number of nozzles in the tank. 170kg, and 3000kg models, the number of nozzles was 4,8. I gave the points to confirm the activation of the powder. This study describes the optimal conditions for activation depending on the type of dry chemical powder tank and the number of nozzles.

1. Introduction

According to the "Marine Accident Statistics" at the Maritime Tribunal, the number of ship fires and explosions in the Republic of Korea is 465 cases (2011 ~ 2015) for 5 years, accounting for about 6% of all marine casualties.
Among them, the oil tanker fire and explosion accident accounted for 8% of the total of 256 cases[1].
Generally, half of ship fire accidents are reported to occur in the engine room. In case of an accident, it is difficult to receive support for fire suppression from the land, and therefore, many lives and property are lost[2].
The Powder fire extinguishing system is a fire extinguishing system that suppresses fire which is less effective by water extinguishing method such as sprinkling water or scattered water and combustible gas or oil with the risk of fire spreading.
The powder fire extinguishing system is a device that injects N2 gas filled in N2 cylinder into powder tank, push powder fire extinguishing agent into N2 gas pressure in powder tank and spray fire extinguishing agent in fire area.
The principle of extinguishing the powder absorbs heat from the fire by the rapid endothermic reaction of the powder which is released in the place where the fire occurs.
As the volume of the non-combustible powder expands, the powder covers the fire and blocks the supply of oxygen, thereby suppressing the fire.
The powder fire extinguishing system is suitable for the extinguishing of large-scale flammable gas fires and is suitable for the fire extinguishing of liquefied natural gas carrier (LNGC), liquefied petroleum gas carrier (LPGC), liquefied natural gas - floating production storage facility (FLNG)), Drillships, and gas and marine plant facilities.

Powder in Dry Powder Tank mixed well with N2 gas in case of fire suppression needs to be maintained after optimal activation.
In this study, we have studied the activation of powders in accordance with the supply of N2 gas in the dry powder tank and studied how the powder is activated according to the number of nozzles in the dry powder tankAnother section of your paper

2. Equations and mathematics

The volume fraction equation, turbulent kinetic energy and flow equation applied to the calculation are as follows[3-5].
The volume fraction equation is as follows.

$$\frac{\alpha_q^{n+1}\rho_q^{n+1}-\alpha_q^n\rho_q^n}{\Delta t}V+\sum_f(\rho_q^{n+1}U_f^{n+1}\alpha_{q,f}^{n+1})=[s_{\alpha_q}+\sum_{p=1}^{n}(\dot{m}_{pq}-\dot{m}_{qp})]V \quad (1)$$

The turbulent kinetic energy k is as follows.

$$\frac{\partial(\rho k)}{\partial t}+\frac{\partial(\rho k u_{i})}{\partial x_i}=\frac{\partial}{\partial x_j}\left[\frac{\mu_t}{\sigma_k}\frac{\partial k}{\partial x_j}\right]+2\mu_t E_{ij}E_{ij}-\rho\epsilon \quad (2)$$

3. Calculation condition

3.1. Preparation condition

The fluid analysis and the activation analysis of the powder were carried out for 170kg and 3000kg dry powder tanks. The number of nozzles was set to 4 and 8 for each model. The inlet consists of two pipes, and the outlet consists of one pipe(Figure1). Two thin pipes come in with nitrogen gas and nozzles at their ends. And the activated powder is discharged into a thick pipe.

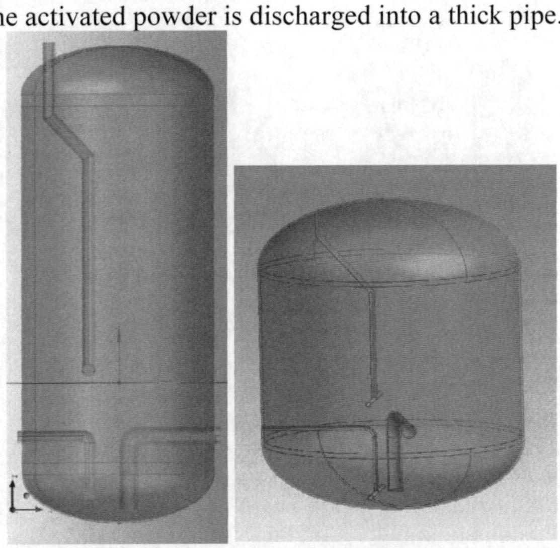

Figure 1. The image of dry powder tanks (170, 3000kg)

3.2. Calculation condition

Before calculation, boundary condtion was set as shown in Table 1. At first, type of fluid inlet was set as Pressure inlet and the value of the fluid pressure was set at 15.7 bar.abs. The properties of the powder are shown in Table 2.

Table 1. Cases of the powder tank

	Powder Tank	Nozzle	Outlet
Case 1	170kg	4	None
Case 2	170kg	8	None
Case 3	3000kg	4	None
Case 4	3000kg	8	None

Table 2. Properties of the powder

Items	Requirements
Main Components	Potassium Bicarbonate
Type	Pure substance
Group	Particle Solid
Density	0.95 g/cm^3
Particle distribution	<40μm : 55% 40 – 63μm : 20 ± 8% 63 - 125μm : 20 ± 8% >125μm : 5 ± 5%

4. Result and discussion

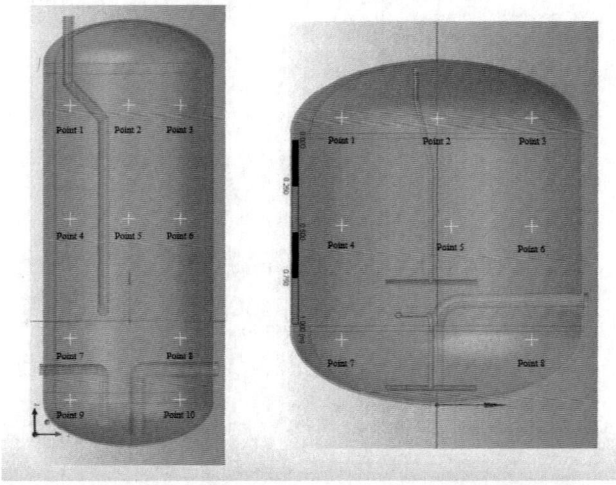

Figure 2. Set the points of dry powder tank

4.1. Result

We set the points as shown in Figure 2 to confirm the activation of the powder. We set 10 points for 170kg and 8 points for 3000kg. In this study, we proceed to analyze for 20 seconds to see that the powder is activated. And it was interpreted as a condition without outlet[6].

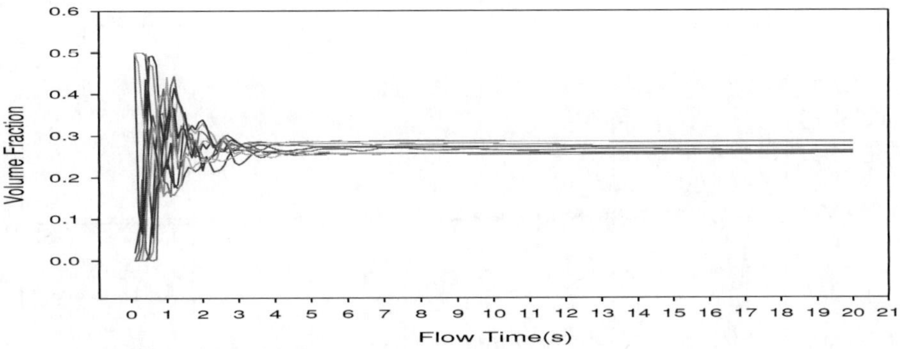

Figure 3. Powder volume fraction of case 1

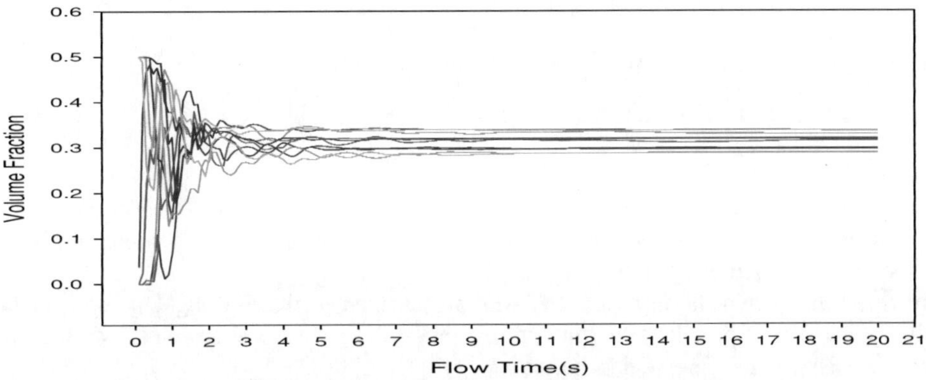

Figure 4. Powder volume fraction of case 2

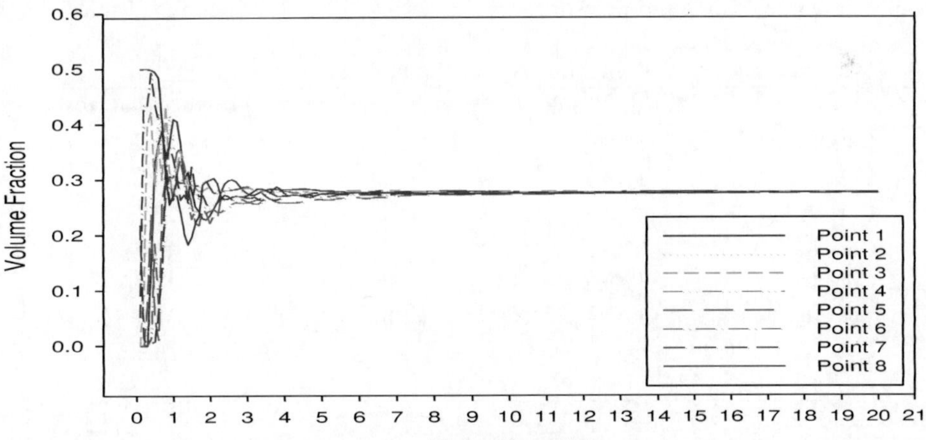

Figure 5. Powder volume fraction of case 3

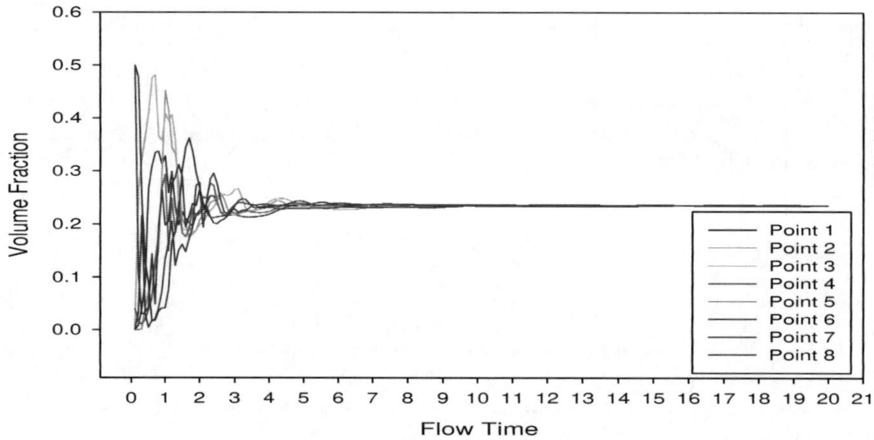

Figure 6. Powder volume fraction of case 4

Figure 3-6 shows the powder activation in case 1-4.

4.2. Discussion

When nitrogen gas was pressurized in a 170kg-Dry powder tank for 20 seconds, some mixing appeared but it was not completely mixed. In the 170kg-Dry powder tank, when the number of nozzles is 4, the dispersion of powder is better.

It was confirmed that when the nitrogen gas was pressurized in the 3000kg-Dry powder tank for 20 seconds, the dispersion of powder was uniform. In the 3000kg-Dry powder tank, when the number of nozzles is 8, the dispersion of powder is better.

As a result, it was confirmed that the powder was well dispersed in case of 3000kg, but in case of 170kg, both cases showed insufficient activation. Therefore, we reanalyzed the powder to activate the 170kg model. The value of the fluid pressure was set at 16.7, 17.7, 18.7 bar.abs (Table 3).

Table 3. Cases of the powder tank

	Powder Tank	Nozzle	Pressure (bar.abs)
Case 5	170kg	4	16.7
Case 6	170kg	4	17.7
Case 7	170kg	4	18.7
Case 8	170kg	8	16.7
Case 9	170kg	8	17.7
Case 10	170kg	8	18.7

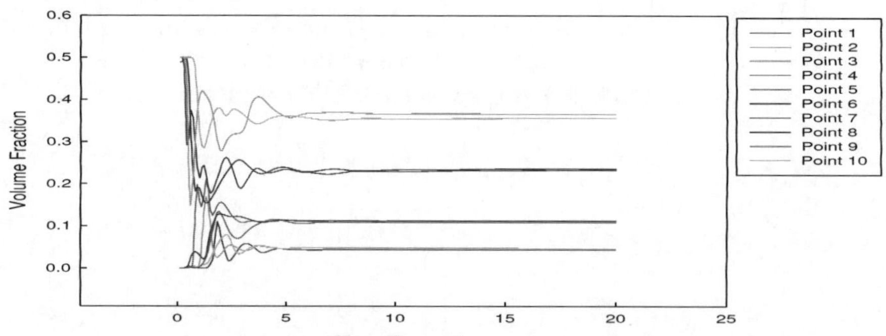

Figure 7. Powder volume fraction of case 5

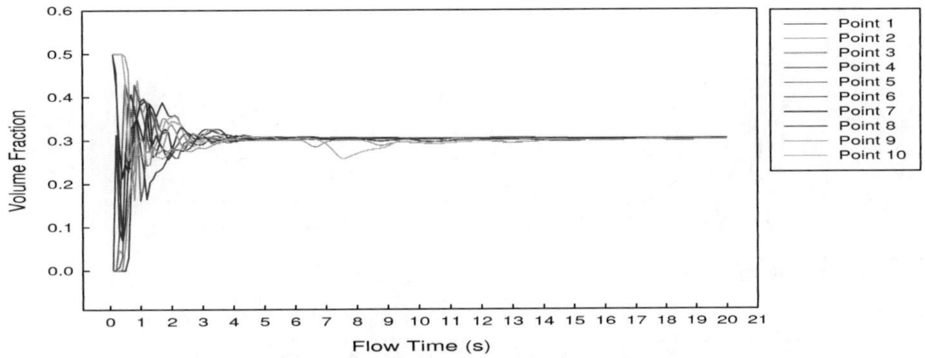

Figure 8. Powder volume fraction of case 6

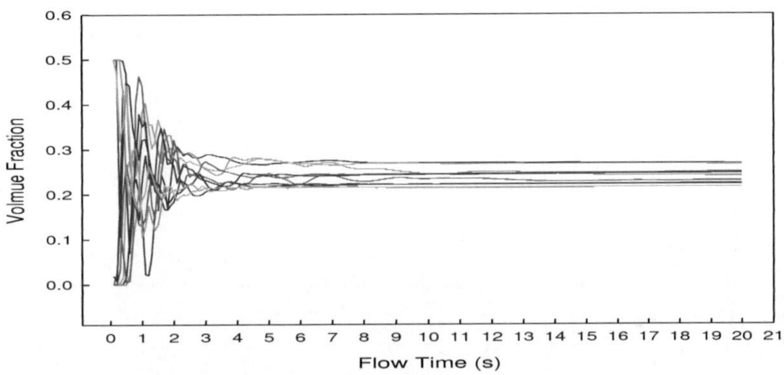

Figure 9. Powder volume fraction of case 7

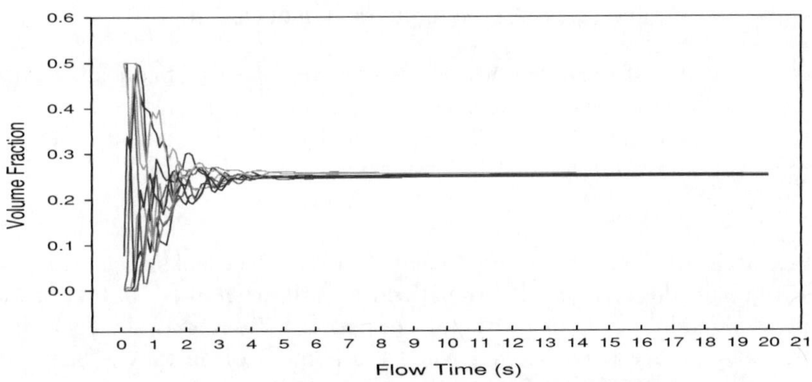

Figure 10. Powder volume fraction of case 8

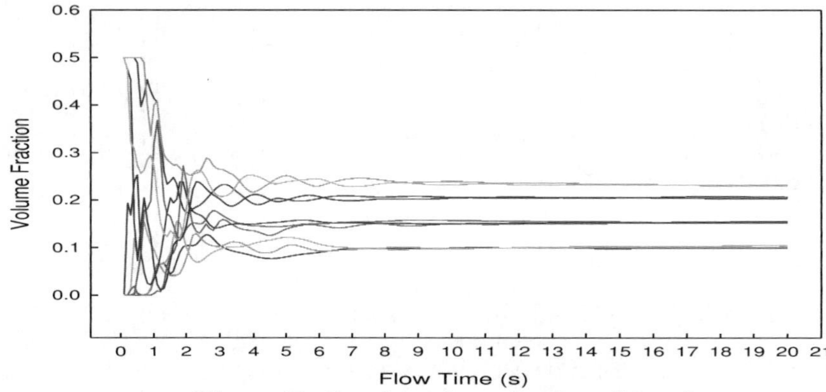

Figure 11. Powder volume fraction of case 9

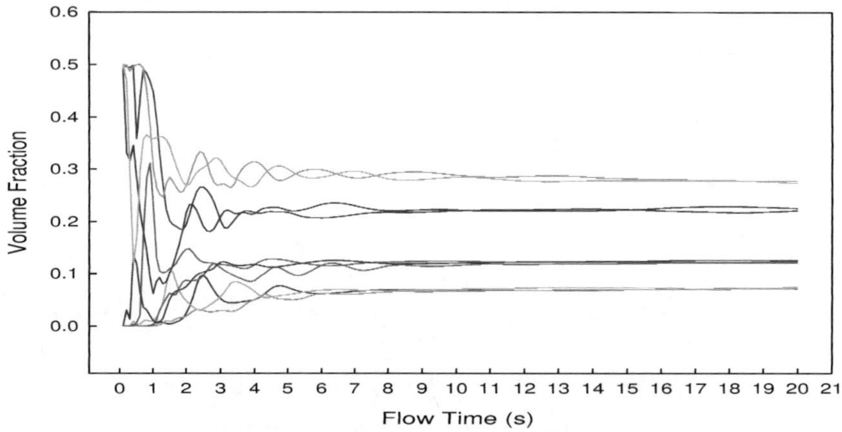

Figure 12. Powder volume fraction of case 10

As a result of analysis, it was confirmed that powder was activated at 17.7 bar.abs when the number of nozzles was 4 when more pressure was applied. When the number of nozzles was 8, the powder was activated well at 16.7 bar.abs. In the case of 170kg-Dry powder tank, the number of nozzles was 8 and the highest powder activation was observed at 16.7 bar.abs.

5. Conclusion

In this study, we investigated the activation of powders in dry powder tanks. The effect of the number of nozzles and pressure on the powder is summarized as follows.It was confirmed that when the nitrogen gas was pressurized in the 3000kg-Dry powder tank for 20 seconds, the dispersion of powder was uniform. In the 3000kg-Dry powder tank, when the number of nozzles is 8, the dispersion of powder is better. In the case of 170kg model, it was confirmed that there is a pressure to activate the powder according to the number of nozzles. When the number of nozzles was 4 and 8, it was activated well at 17.7 bar and 16.7 bar.It was defined that the state of activation depends on the shape of powder tank of 170kg and 3000kg. Therefore, the more the shape of the tank is closer to the ellipse, the better the activation is.

6. References

[1] Dong-Suk Kim, "Development of fire extinguishing system suitable for unmanned engine room f a small ship", T. of Korean Institude of Fire Sci. & Eng., Vol. 20, No. 3, 2006.
[2] Ha-Sung Kong and Jong-Jin Kim, "A study on improving extinguishing capacity of mobile water mist equipment", Korea Safety Management&Science, Vol. 11, No. 1, pp 15-26, 2009.

[3] K. W. Chu, B. Wang and A. B. Yu, "DISCRETE PARTICLE SIMULATION OF THE GAS-SOLID FLOW IN A CIRCULATING FLUIDIZED BED," Engineering Conferences International, 2007.
[4] Chan-Young Cho, "Numerical Study on rotational flow and slurry dispersion in stirred tanks," unpublished.
[5] "Fluent User Guide," unpublished.
[6] Myeong-Rok Ryu and Kweon-Ha Park, "Hydrogen explosion effects at a containment building following a severe accident", Journal of the Korean Society of Marine Engineering, Vol. 40, No. 3 pp. 164~172, 2016.

Acknowledgment
This research was supported by creative convergence R&D of local main industry promotion business.

IOP Conf. Series: Materials Science and Engineering **239** (2017) 012017 doi:10.1088/1757-899X/239/1/012017

Investigation of phase transformation of CoNi alloy nanowires at high potential

A Mukhtar[1], T Mehmood[1] and K M Wu[1]

[1]The state key laboratory of Refractories and Metallurgy, Hubei Collaborative Innovation Center for Advanced Steels, International Research Institute for Steel Technology, Wuhan University of Science and Technology, Wuhan,P. R. China
Email: aleeza.mukhtar@yahoo.com

Abstract. Among various ferromagnetic materials, Cobalt Nickel (CoNi) is considered as an important alloy because of its enhanced magnetic and catalytic properties as well as better mechanical properties as compared to single Co nanowires. Fabrication of binary alloy nanowires using electrochemical deposition is cheap and challenging due to the different synthesis conditions of magnetic materials being used. In this study phase transformation of CoNi alloy nanowires are observed using Direct current (DC) deposition technique with in Anodic Aluminium Oxide (AAO) template. The average pore diameter of the CoNi alloy nanowires was approximately 60 nm which is equal to the diameter of the nanopore. Results show that at a high potential of –3.5 V fcc CoNi alloy nanowires and low potential of –1.0V hcp CoNi alloy nanowires are formed. The phase transformation of CoNi alloy nanowires at high potential can be attributed to, a small number of large nuclei are formed at the high deposition potential.

1. Introduction

Co and Ni alloy phase diagram shows that [1], Co and Ni exhibit complete solid solution in the fcc phase at temperatures between the solidus and the allotropic transformation temperature, alloy phase diagram confirms that transformation occurs at the temperature about 300 ^{0}C[1]. Therefore, it is known that fcc $Co_{84.43}Ni_{15.51}$ is a high-temperature phase.

Our previous research articles show that structure of Co depends on deposition parameters. High potential and electrolyte concentration favors fcc Co phase while low potential and electrolyte concentration favors hcp Co phase [2, 3]. The recent article proposed a new model which describes the formation of fcc Co with DC deposition at the high potential of -3.0V [3].

A lot of research work has been done on CoNi alloy nanowires regarding their chemical composition and structure in AAO template [4-6]. Many researchers [7, 8] found the mixed phase of hcp and fcc at 30:70 nickel cobalt ion concentration in an electrolytic solution using AC Electrodeposition. However, no work has been done on the effect of potential on phase transformation of CoNi alloy nanowires. The current research attempts to investigate the effect of potential on the crystal structure and composition of Co-Ni alloy nanowires. Hope that this research would be helpful to compose other ferromagnetic alloy nanowires using AAO template in future.

2. Material and methods

The detailed experimental procedure for preparation of the AAO template is clearly explained in our previous research article [2]. The electrolyte was $0.24MCoSO_4 \cdot 7H_2O + 0.06MNiSO_4 \cdot 6H_2O + 0.65MH_3BO_3$ .The *pH* of electrolyte was adjusted to 2.5 by adding the 1M H_2SO_4 solution. The Potentiostatic deposition was conducted in a three-electrode cell at the bath temperature of 25 ^{0}C. AAO template area for depositing CoNi alloy nanowires was 0.608cm^2. The graphite was used as the counter electrode (area of 14.7cm^2) and saturated calomel (SCE) was used as reference electrode. The structure and morphology of CoNi alloy nanowires were investigated by using XRD (X'Pert PRO MRD, PANalytical, Netherlands) with CuK$_\alpha$ radiation, FE-SEM (NOVA 400 Nano) and EDS (Le350 PentaFETx-3). For SEM observations AAO template was partly dissolved with 5 wt% NaOH solution, and then carefully rinsed with deionized water for several times.

3. Results

Fig.1a-b shows SEM and EDX images of CoNi alloy nanowires -1.0V and, Fig.2 a-b shows SEM EDX images of the CoNi alloy nanowires deposited at –3.5V.

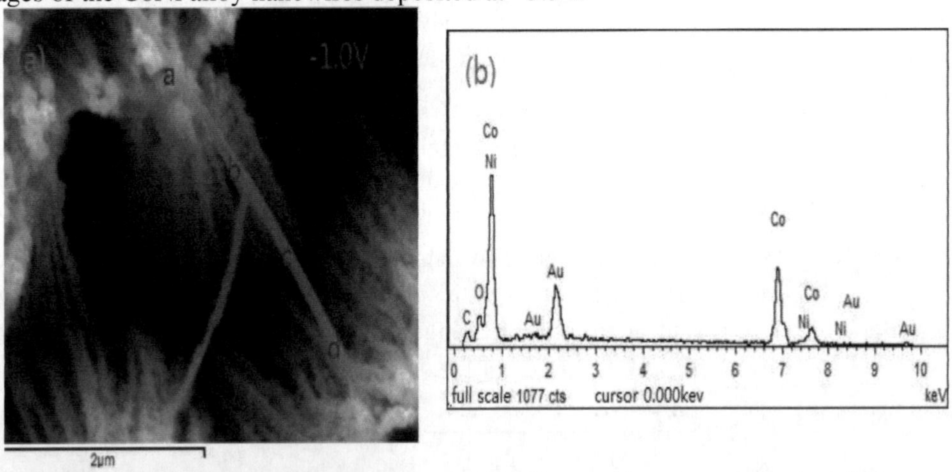

Figure.1 (a) SEM (b) EDX images of CoNi nanowires deposited at –1.0 V

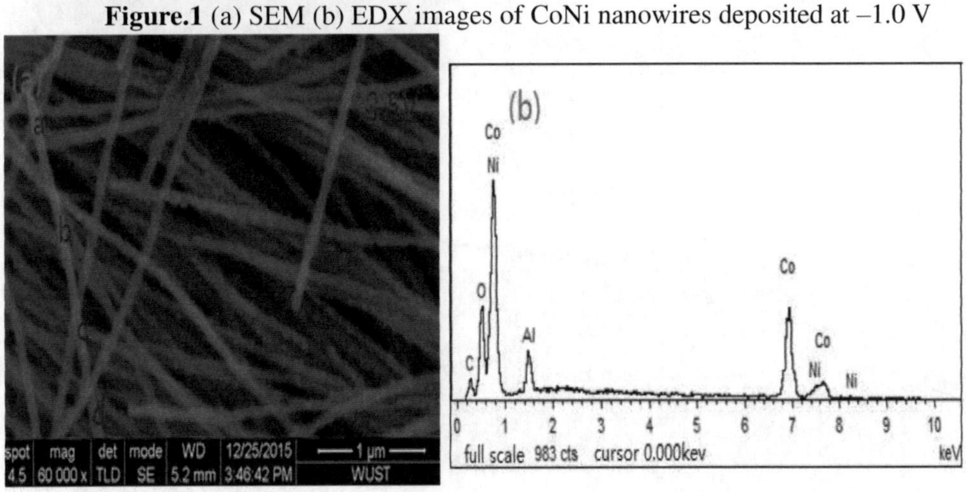

Figure.2 (a) SEM (b) EDX images of CoNi nanowires deposited at –3.5 V.

The compositions of Co and Ni were measured using EDX at points a, b, c, d. The averaged compositions of Co and Ni at -1.0V are 80.60% and 19.39%, averaged compositions of Co and Ni at -3.5V are 84.43% and 15.51% as shown in table.1. It is clear from table 1 that, high potential increases the content of Co in the CoNi alloy nanowires.

IOP Conf. Series: Materials Science and Engineering **239** (2017) 012017 doi:10.1088/1757-899X/239/1/012017

Table 1. The composition of Co and Ni in CoNi alloy nanowires deposited at -1.0V and -3.5V

-1.0V		-3.5V	
Co (at.%)	Ni(at.%)	Co (at.%)	Ni (at.%)
80.64	19.33	84.6	15.38
80.45	19.29	83.90	15.88
80.33	19.33	84.55	15.23
81.01	19.64	84.68	15.56

Fig. 3(a-b) shows the XRD patterns of CoNi alloy nanowires deposited at –1.0 V and –3.5 V. The peak at $2\theta = 41.6^{\circ}$ is well fitted to hcp $Co(10\bar{1}0)$ (d = 2.165 Å) . The CoNi phase diagram [1] shows that for $Co_{80.60}Ni_{19.39}$, Co and Ni exhibit complete solid solution in the hcp phase below the transformation temperature. Furthermore, the atomic radius of Co and Ni is the same, which is 1.25 Å [9] and after dissolving Ni atoms into Co lattice sites, Ni atoms cannot change the spacing (d) of planes. Hence, the peak at $2\theta = 41.6^{\circ}$ is unambiguously attributed to the diffraction of $(10\bar{1}0)$ plane of hcp alloy phase $Co_{80.60}Ni_{19.39}$. The peak at $2\theta = 76^{\circ}$ (d = 1.2520 Å) can be fitted to fcc Co (220) (d =1.253 Å) according to file No.15806. It is clear from Fig.3 at high potential of -3.5V fcc CoNi alloy nanowires are formed.

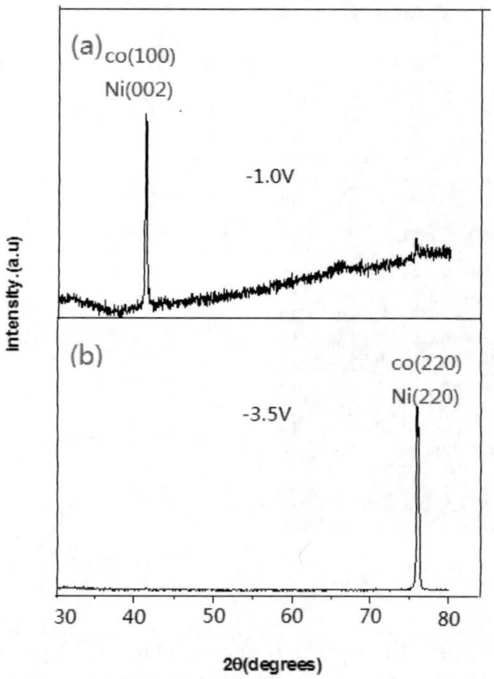

Figure.3 XRD of CoNi alloy nanowires deposited at (a)–1.0V (b) –3.5 V

4. Discussion

When the negative potential is applied to the working electrode which is AAO template, divalent cobalt and nickel ions which are surrounded by hydration shells move towards the cathode, get reduced and neutralized,

$$Co^{+2} + 2e^{-} \rightarrow Co \quad (1)$$

$$Ni^{+2} + 2e^{-} \rightarrow Ni \quad (2)$$

The attractive interaction between water molecule and neutral Co and Ni ions is zero, so hydrated water molecules are displaced and neutral metal atom gives an intermetallic phase according to the following chemical reaction

$$Co + Ni \rightarrow CoNi \quad (3)$$

Than cobalt and nickel diffuses to surface sites where it incorporates into the metal lattice, after the incorporation into the metal lattice the CoNi alloy nanowires are formed [10].

The total free energy of a crystal nucleus is the sum of the bulk free energy and surface free energy of the nucleus. The bulk free energy of hcp Co is lower than that of fcc Co at temperatures below 420 ^{0}C [11]. However, it is well known that the surface energy of fcc Co is lower than that of hcp Co; previous researchers [3, 11, 12] use the surface energy effect to explain the experimental observations: smaller Co grains favor fcc structure and larger Co grains favor hcp structure. For alloy nanowires studied here, the bulk free energy of hcp phase is lower than of fcc phase at room temperature. However, the surface energy of hcp phase can be higher than of fcc phase, based on two reasons. (i), high content of Co found inside alloy nanowires than used in electrolyte concentration. (ii), the hcp $(10\bar{1}0)$ and fcc (110) are favored growth surfaces. Surface energies of hcp $(10\bar{1}0)$ and fcc (110) are not calculated yet. Though, one can find that the surface energy of fcc alloy (110) plane is lower than that of the hcp alloy $(10\bar{1}0)$ plane. According to their atomic radius [9] and the surface structure of hcp $(10\bar{1}0)$ and fcc (110) [13], the surface atom density of hcp $(10\bar{1}0)$ and fcc (110) is calculated to be $9.8\text{x}10^{14}$ cm^{-2} and $1.13\text{x}10^{15}$ cm^{-2}. The surface energy depends on the surface atom density: the surface with lower atom density is usually the surface with higher surface energy[14] and thus the surface energy of hcp $(10\bar{1}0)$ is higher than that of fcc (110). Hence, we suggest that for small nucleus size, an fcc CoNi nucleus is thermodynamically more stable than an hcp CoNi nucleus due to the significant surface energy effect.

The size of critical cluster formed in deposition, N_c is [3, 15]

$$N_c = \left(\frac{2b}{3ze\eta}\right)^3 \quad (4)$$

where z is the valence of hydrated metal ions, e is the electron charge, η are the overpotential and b depends on the geometrical shape of the cluster. It is obvious from Eq. (4) that at high the potential, small critical clusters are formed, favoring the formation of fcc phase. While lower potential leads to the formation of larger critical clusters, thus favoring the formation of hcp phase. Therefore, the $Co_{84.43}Ni_{15.51}$ alloy nanowires formed at –3.5 V has fcc structure, and $Co_{80.60}Ni_{19.39}$ alloy nanowires formed at –1.0 V has hcp structure.

The electrochemical deposition of an alloy is the co-deposition of two metals. The compositions of alloy depend on the polarization curves of depositing these two metals [15]. Our previous studies [10] show that the polarization curves of Co and Ni nanowires is linear and the slope of the polarization curve of depositing Co is larger than of depositing Ni, indicating that the rate of deposing Co atoms increases more quickly with increasing potential than that of depositing Ni atoms. Therefore high potential increases the content of Co inside CoNi alloy nanowires.

Conclusion

The present study shows that the AAO template with periodic hexagonally arranged nano-pores with in well aligned CoNi nanowires using different potentials is successfully fabricated. Results show that the Phase transformation of CoNi alloy nanowires occurs at the high potential of –3.5 V. The high potential fcc phase is a metastable phase, while the CoNi alloy nanowires deposited at low potential has hcp phase. The formation of fcc alloy nanowires can be attributed to smaller critical clusters formed at the high potential.

References

[1] Nishizawa T, Ishida K 1983 The Co–Ni (Cobalt-Nickel) system *Bulletin of Alloy Phase Diagrams* **4** 390 .

[2] Mukhtar A, Mehmood T, Khan BS, Tan M. 2016 Effect of Co2+ concentration on the crystal structure of electrodeposited Co nanowires *J. Cryst. Growth.* **441** 26-32.

[3] Mehmood T, Shahzad Khan B, Mukhtar A, Chen X, Yi P, Tan M 2014 Mechanism for formation of fcc-cobalt nanowires in electrodeposition at ambient temperature *Mater. Lett* **130**:256-8.

[4] Ghahremaninezhad A, Dolati A. 2009 A study on electrochemical growth behavior of the Co–Ni alloy nanowires in anodic aluminum oxide template *J. Alloys Compd* **480** 275.

[5] Thongmee S, Pang HL, Yi JB, Ding J, Lin JY, Van LH. 2009 Unique nanostructures in NiCo alloy nanowires *Acta Mater* **57** 2482.

[6] Kwag Y-G, Ha J-K, Kim H-S, Cho H-J, Cho K-K. 2014 Co–Ni Alloy Nanowires Prepared by Anodic Aluminum Oxide Template via Electrochemical Deposition *Journal of Nanoscience and Nanotechnology* **14** 8930.

[7] Ali G, Maqbool M. 2013Fabrication of cobalt-nickel binary nanowires in a highly ordered alumina template via AC electrodeposition *Nanoscale Research Letters* **8** 352.

[8] Kawamori M, Yagi S, Matsubaraa E. 2012 Nickel alloying effect on formation of cobalt nanoparticles and nanowires via electroless deposition under a magnetic field *J Electrochem Soc* **159**(2) E37.

[9] Kittel C 2004 *Introduction to Solid State Physics* (New York: Wiley).

[10] Shahzad Khan B, Mehmood T, Mukhtar A, Tan M 2014 Effect of workfunction on the growth of electrodeposited Cu, Ni and Co nanowires *Mater. Lett* **137** 13.

[11] Sato H, Kitakami O, Sakurai T, Shimada Y, Otani Y, Fukamichi K. 1997 Structure and magnetism of hcp-Co fine particles *J. Appl. Phys* **81** 1858.

[12] Owen E, Jones DM. 1954 Effect of grain size on the crystal structure of cobalt. *Proceedings of the Physical Society Section B* **67** 456.

[13] Tan M, Chen X. 2012 Growth Mechanism of Single Crystal Nanowires of fcc Metals (Ag, Cu, Ni) and hcp Metal (Co) Electrodeposited *J Electrochem Soc* **159**:K15.

[14] Nix RM 2003 *An introduction to surface chemistry* (School of biological and chemical sciences Queen Marry, University of london)

[15] Paunovic M, Schlesinger M 1998 *Fundamentals of Electrochemical Deposition* (New York: Wiley).